Exploratory Examples for Real Analysis

Exploratory Examples for Real Analysis

Joanne E. Snow
Saint Mary's College

Kirk E. Weller
University of North Texas

Published and Distributed by
The Mathematical Association of America

CLASSROOM RESOURCE MATERIALS

Classroom Resource Materials is intended to provide supplementary classroom material for students—laboratory exercises, projects, historical information, textbooks with unusual approaches for presenting mathematical ideas, career information, etc.

101 Careers in Mathematics, 2nd edition edited by Andrew Sterrett

Archimedes: What Did He Do Besides Cry Eureka?, Sherman Stein

Calculus Mysteries and Thrillers, R. Grant Woods

Combinatorics: A Problem Oriented Approach, Daniel A. Marcus

Conjecture and Proof, Miklós Laczkovich

A Course in Mathematical Modeling, Douglas Mooney and Randall Swift

Cryptological Mathematics, Robert Edward Lewand

Elementary Mathematical Models, Dan Kalman

Environmental Mathematics in the Classroom, edited by B. A. Fusaro and P. C. Kenschaft

Essentials of Mathematics, Margie Hale

Exploratory Examples for Real Analysis, Joanne E. Snow and Kirk E. Weller

Geometry From Africa: Mathematical and Educational Explorations, Paulus Gerdes

Identification Numbers and Check Digit Schemes, Joseph Kirtland

Interdisciplinary Lively Application Projects, edited by Chris Arney

Inverse Problems: Activities for Undergraduates, Charles W. Groetsch

Laboratory Experiences in Group Theory, Ellen Maycock Parker

Learn from the Masters, Frank Swetz, John Fauvel, Otto Bekken, Bengt Johansson, and Victor Katz

Mathematical Evolutions, edited by Abe Shenitzer and John Stillwell

Mathematical Modeling in the Environment, Charles Hadlock

Mathematics for Business Decisions Part 1: Probability and Simulation (electronic textbook), Richard B. Thompson and Christopher G. Lamoureux

Mathematics for Business Decisions Part 2: Calculus and Optimization (electronic textbook), Richard B. Thompson and Christopher G. Lamoureux

Ordinary Differential Equations: A Brief Eclectic Tour, David A. Sánchez

Oval Track and Other Permutation Puzzles, John O. Kiltinen

A Primer of Abstract Mathematics, Robert B. Ash

Proofs Without Words, Roger B. Nelsen

Proofs Without Words II, Roger B. Nelsen

A Radical Approach to Real Analysis, David M. Bressoud

She Does Math!, edited by Marla Parker

Solve This: Math Activities for Students and Clubs, James S. Tanton

Student Manual for Mathematics for Business Decisions Part 1: Probability and Simulation, David Williamson, Marilou Mendel, Julie Tarr, and Deborah Yoklic

Student Manual for Mathematics for Business Decisions Part 2: Calculus and Optimization, David Williamson, Marilou Mendel, Julie Tarr, and Deborah Yoklic

Teaching Statistics Using Baseball, Jim Albert

Writing Projects for Mathematics Courses: Crushed Clowns, Cars, and Coffee to Go, Annalisa Crannell, Gavin LaRose, Thomas Ratliff, Elyn Rykken

MAA Service Center
P.O. Box 91112
Washington, DC 20090-1112
1-800-331-1MAA   FAX: 1-301-206-9789

Preface

"In mathematics too some things appear to be not easy to prove for want of a definition
... But, when the definition is expressed, the said property is immediately manifest."

— Aristotle

The limit is the essential idea of the calculus. It took more than 2000 years for the definition to evolve into the form we have today. The long struggle to formulate clearly and accurately this definition testifies to its richness and complexity. Thus, it is no surprise that students have difficulty understanding the limit and related notions. These laboratory exercises represent our efforts to help students become comfortable with the definition of limit and other important, related definitions in analysis, such as supremum and infimum, boundedness, limit superior and limit inferior, continuity, and uniform convergence.

This list does not come close to exhausting the list of topics covered in a first course. That was not our intention. Rather, the labs are designed to help students develop useful conceptions of the most elementary definitions and tools so that they can apply these fundamental ideas in the study of more difficult and involved topics such as differentiation, integration, and series of functions. In using the labs with our students, we have found that they have become more adept at solving problems and in proving theorems that involve the application of these fundamental concepts. In this sense, the labs have succeeded in helping our students to see the definitions of supremum/infimum, boundedness, limit, and continuity as building blocks, as opposed to stumbling blocks.

In designing these labs, we were inspired by the work of Ellen Parker of DePauw University, who has used a laboratory approach to teach group theory. Each of her labs was designed to facilitate concept development or to guide students to formulate conjectures on the basis of working with a vast collection of finite groups that could be easily accessed and studied through use of a computer software package. Her approach intrigued us because it reflects the process by which some mathematicians formulate ideas. We have tried to emulate the same approach in this project, although our materials do not depend upon the use of a particular software package. In fact, several of the labs are designed to be done by hand, and many can be completed using a graphing calculator. For those labs where the use of technology proves to be beneficial, we have written Visual Guide Sheets using *Maple* code. Each Visual Guide can be found by going to the

Real Analysis Resource link at www.saintmarys.edu/~jsnow. The Visual Guides also appear in the appendix. They will be updated periodically online to make refinements and additions.

One important way mathematicians understand concepts and suggest theorems is by working with specific examples. As a means of encouraging this practice, we have students work with many examples. However, we are not advocating a position that mathematics is an empirical field of inquiry. Although often presented to students as a finished product, the development of mathematical ideas is an often long and arduous process that involves work with many examples. We want to give our students a sense of this. Another goal is to help students understand exactly what is guaranteed by a definition, so that they neither read more into what a definition promises nor miss what is asserted.

Recognizing how a particular example meets or fails to meet a definition gives one power over that definition. As students learn fundamental definitions, we have targeted the development of certain skills. For instance, we want students to be able to read definitions accurately. In several of the labs, the students are required to formulate their own versions of a definition based on their study of the examples that have been presented. We then have them apply their definitions to see whether their conceptions need to be modified. Several of the definitions, particularly those of limit and continuity, involve the use of quantifiers. In order to help students distinguish between universal and existential quantifiers and the importance of their placement in the statement of a definition, we have students consider the consequences of replacing an existential quantifier with a universal quantifier, or vice-versa. We also ask students to write proofs in which they have to apply the definitions they have learned. This prepares them to deal more effectively with complex situations that arise later in the course. The examples we have chosen to motivate development of these concepts are familiar, fairly standard, or both. This was done intentionally, as the purpose of these materials is to help students to develop a strong, basic sense of the fundamental ideas in analysis.

How these skills are practiced varies according to the purpose of the individual lab. Students may be asked to formulate definitions, make connections between different concepts, derive conjectures, or complete a sequence of guided tasks designed to facilitate concept acquisition. Each lab has three basic components: making observations and generating ideas from hands-on work with examples; thinking critically about the examples by recognizing patterns, connecting ideas, and drawing conclusions; and answering additional questions for reflection. We include these additional questions to challenge students; often these are harder questions, leading to a deeper level of the topic at hand or foreshadowing a related topic. We usually have the students work on each lab in groups of two to four students. If they don't complete the lab during the class period, they are required to arrange to meet together to finish their work outside of class. They are then required to prepare a written report that includes the results of their investigation. To help in assigning lab reports, sample report guides can be found at the Real Analysis Resource link at www.saintmarys.edu/~jsnow.

The written reports serve multiple purposes — for the students and the instructor. Having the ability to verbalize what has been learned indicates a level of mastery of the topic. The reports help students with this aspect of learning. These exercises in writing also help students with their proof-writing skill. The reports help the instructor assess students' success with the labs. By identifying misconceptions, particular points of difficulty, and areas that need further emphasis, the labs can be used to help an instructor to prepare lecture and discussion sessions. To grade these reports, we have used the following performance criteria: accuracy of responses, accuracy in the use of mathematical notation and terms, clarity, completeness, and precision in responses and explanations.

When we started this project, we were concerned about content coverage. Specifically, would use of the labs limit the number of topics we could cover? Would their implementation result in students being poorly prepared for graduate school or for other upper division courses? We have used various versions of this manuscript with six different groups of analysis students of varying abilities at two different institutions: a highly selective, single gender, regional liberal arts college where some students pursue graduate study, and a minimally selective, co-educational, regional liberal arts college consisting of a large proportion of first-generation college students who generally do not pursue graduate study. For each instantiation, the labs have been used as a supplement to *Elementary Analysis: The Theory of Calculus*, by Ross (Springer). In both institutions, analysis is offered in a two-semester format in which the following topics are covered: properties of the real numbers, sequences, continuity, sequences and series of functions, differentiation, and integration.

Our use of the labs did not reduce content coverage. We employed a variety of strategies to ensure this. Rather than introduce a topic by presenting a lecture, students are introduced to the topic as they work through one of the labs. As they prepare their report, they are encouraged to make note of questions that arise and difficulties they face, and to consult related portions of the text to see how ideas are related. The class period that follows can then be used to answer their questions, to discuss their misconceptions and to resolve their difficulties, and to check whether they have, or are making, the necessary connections. The conceptions and ideas that are generated through this cycle — lab period, report preparation (done outside of class), class discussion — can be reinforced by assigning homework from the text and/or the Questions for Reflection. Another possible approach is to introduce a topic by giving a short presentation, followed by assignment of the related lab, after which a formal lecture is delivered. In this case, the introductory lecture need not be as detailed as that of an ordinary introduction to the topic, and the subsequent lecture can be more focused on tying up loose ends. The labs could also be used as part of a lecture. In this case, a lab could be divided into a series of in-class activities that could be used in the place of simply presenting examples. In our experience, we have generally discovered that we need to devote at least one 50-minute class period for each lab to provide students the opportunity to ask questions, to identify particular areas of difficulty, and to get an overall sense of the issues being addressed. There are three exceptions to this recommendation: Lab 6, Lab 10, and Lab 12. These are fairly lengthy labs that generally require two 50-minute class periods to complete. However, if students have regular access to a computer lab, one 50-minute period could be allotted, with the balance to be completed outside of class.

Several of the labs do make reference to one another. However, only in the cases of Labs 4 and 6 and Labs 10 and 11 is the dependence non-trivial. This means that the manuscript can be used in whatever way best suits an instructor's particular course design. For instance, for a unit on continuity and limit, one could use Labs 9, 10, and 11. To introduce notions of boundedness, an instructor might elect to use Lab 1. There are three other 'stand-alone' labs: Lab 2, which considers the definitions of supremum and infimum, Lab 8, which introduces the notions of limit superior and limit inferior, and Lab 12, which focuses on pointwise and uniform convergence. There are two sets of labs on sequences, one dealing with issues related to formulation, application, and negation of the definition (Labs 3, 4, 5), and the other set that deals with properties of sequences (Labs 6, 7). Although we have used the manuscript with the Ross text, all of the notation is standard, and there is no reference to this particular text.

Use of these materials appears to have enhanced our students' understanding of the definitions of supremum/infimum, convergence of a sequence, limit superior and inferior, limit of a function, and continuity. When compared with prior analysis classes in which we did not use the labs, our anecdotal evidence suggests that groups of students who used the labs generally wrote better proofs and demonstrated greater mastery of fundamental concepts. Use of these materials has also appeared to facilitate a more inquisitive, positive classroom environment. Both of these observations have been particularly evident with students for whom analysis was their first upper division course or for those for whom mastery of course concepts was more difficult. We believe the labs would be most effective in one of the following scenarios:

- A class of fairly average students for whom analysis is their first upper division course.

- A group of students with a wide range of abilities for whom a cooperative approach focusing upon fundamental concepts could help to close the gap in skill development and concept acquisition.

- An independent study or private tutorial in which the student receives a minimal level of instruction.

- A resource for an instructor developing a cooperative, interactive course that does not involve the use of a standard text.

Given the wide variety of potential uses, we have prepared a solutions manual, which is available at www.saintmarys.edu/~jsnow.

An Overview of the Labs

The first two labs deal with concepts needed to characterize the real numbers: boundedness and supremum/infimum. In Lab 1, students are familiarized with various notions of boundedness: upper and lower bounds, maxima and minima, and suprema and infima. Given sample sets of real numbers, students are asked to determine whether such bounds exist, and, if so, to identify them. They use the information they find in working with the examples to formulate the relationships between the various notions of boundedness. Lab 2 begins with presentation of the "non-epsilon" definition of the supremum of a set. Working with a series of examples, students use this definition to construct the "epsilon" version of the definition of supremum. Students are then asked to show that the two definitions are equivalent.

The focus of Labs 3-5 is the definition of the convergence of a sequence. Given an intuitive characterization of convergence, the goal of Lab 3 is to develop a formal definition of the concept. Students follow a sequence of questions involving graphical and numerical calculations to understand the role of quantifiers, as well as to formulate the "$\epsilon - N$" definition. The emphasis in Lab 4 is on having students use the definition to prove that a given sequence converges to a particular value. For each example, students make a claim on the basis of their intuitions. Then, using graphical and numerical techniques, students are guided through the process of applying the definition. For a given value of ϵ, the students identify a value of N. The process is repeated both to show the dependence of N upon ϵ, as well as to make clear the arbitrariness of ϵ. In the following section, students are introduced to algebraic techniques for constructing a formal proof. The hope is that the prior numerical and graphical work will stimulate understanding of the construction of a formal proof. In Lab 5, students develop the negation of the definition of convergence. For a given sequence, students use the negated definition to show that a specified value cannot serve as the limit. Students use the techniques and insight gained in consideration of a specified value to write general divergence proofs.

The behavior of convergent sequences is the topic of the next two labs. In Lab 6, students consider algebraic combinations (sums, products, and quotients) of sequences and conjecture how the limits of two sequences are related to the limit of the combination. After guiding students through proofs of the conjectures for particular sequences, students are asked to produce proofs of the conjectures for arbitrary sequences. In Lab 7, students determine conditions that yield or are related to convergence. For a variety of sample sequences, students are asked to identify the limit of a sequence if it exists; to determine whether the sequence is bounded; to ascertain monotonicity; and to compute the difference between terms as the values of the index n increases. Students then formulate relationships between convergence and boundedness, convergence and monotonicity, and convergence and the Cauchy criterion.

Lab 8 introduces the notion of limit superior and limit inferior through a series of steps designed to enhance understanding of the formal definition. Students are asked to formulate the relationship between the limit of a sequence, the limit superior, and the limit inferior.

Students explore the topics of continuity and limit in Labs 9-11. In Lab 9, students consider a series of examples using numerical and graphical approaches as a means of devising the "sequence definition" of continuity. In Lab 10, students are provided with the "$\epsilon - \delta$" definition of continuity. They are then asked a series of questions to help them to understand and to apply the definition. In a later section, students are asked to determine what modifications need to be made to the definition of continuity in order to obtain the limit of a function at a point. This provides an opportunity for students to compare and to contrast the two definitions as part of the process of learning to distinguish between the two concepts of limit and continuity. In Lab 11, students gain experience with the "$\epsilon - \delta$" definitions of continuity and limit, as they learn how to use the definitions to prove theorems involving algebraic combinations of functions. Students first work with specific functions, and then modify their work to apply these definitions to arbitrary functions.

In the final lab, students examine pointwise and uniform convergence. In the first part of the lab, students compute pointwise limits for several examples. Students are asked a series of questions about the examples to help them distinguish between simple pointwise convergence and uniform convergence. Based on the results of the data analysis, students are asked to determine when the limit of a sequence of continuous functions is continuous.

We would like to thank Mic Jackson of Earlham College and Zaven Karian, Editor of CRM, as well as the members of the Editorial Board of CRM, for their helpful comments and suggestions as we prepared and revised this project. It is our sincere hope that those who use the labs find them as useful in helping their students to learn the basic elements of real analysis as we have in our collective experience over the course of the last six years.

— Joanne E. Snow and Kirk E. Weller

Contents

1

Boundedness of Sets

1.1 Introduction

In this lab, we examine some properties dealing with boundedness: upper bound, lower bound, supremum, infimum, maximum, and minimum. We apply these various notions to a sample of sets in order to understand clearly the similarities and differences among the terms. Some properties are stronger than others; some imply the existence of one or more of the other properties. As you will see later, these concepts are key in understanding the structure of the real numbers.

The definitions of the relevant terms are presented here. For all the definitions, let A represent a nonempty set of real numbers.

- A real number u is said to be an *upper bound* for a set A if $x \leq u$ for all $x \in A$.

- A real number l is said to be a *lower bound* for a set A if $l \leq x$ for all $x \in A$.

- A set is *bounded* if it possesses both an upper and a lower bound.

- A real number s is the *supremum*, or *least upper bound*, of a set A if s is an upper bound of A and $s \leq u$ for any other upper bound u of A. The supremum is denoted $\sup(A)$.

- A real number t is the *infimum*, or *greatest lower bound*, of a set A if t is a lower bound of A and $t \geq l$ for any other lower bound l of A. The infimum is denoted $\inf(A)$.

- A real number m is the *maximum* of a set A if $m \in A$ and $x \leq m$ for all $x \in A$.

- A real number n is the *minimum* of a set A if $n \in A$ and $n \leq x$ for all $x \in A$.

1.2 Using Examples to Enhance Understanding

First we gather data by examining some sets. Complete the table that follows by providing the desired information about the given sets. In column 1, the set is described. For each set, give an example of the item listed at the head of the column in columns 2–7. Write your answer in the appropriate cell of the table. If the set fails to have one of the items, write DNE in the appropriate cell. In columns 8 and 9, put Y if the property holds for the set and N otherwise.

1

Notation: $\mathbb{R}$ = real numbers, $\mathbb{Z}$ = integers, $\mathbb{N}$ = natural numbers.

Set	Upper bound	Lower bound	Max	Min	Sup	Inf	Is the sup in the set?	Is the set bounded?
1) $\{x \in \mathbb{R} : 0 \le x < 1\}$	1	0	DNE	0	1	0	No	Yes
2) $\{x \in \mathbb{R} : 0 \le x \le 1\}$	1	0	1	0	1	0	Yes	Yes
3) $\{x \in \mathbb{R} : 0 < x < 1\}$	1	0	DNE	DNE	1	0	No	Yes
4) $\{1/n : n \in \mathbb{Z} \setminus \{0\}\}$	1	-1	1	-1	1	-1	Yes	Yes
5) $\{1/n : n \in \mathbb{N}\}$	1	0	1	DNE	1	0	Yes	Yes
6) $\{x \in \mathbb{R} : x < \sqrt{2}\}$	$\sqrt{2}$	DNE	DNE	DNE	$\sqrt{2}$	DNE	No	No
7) $\{1, 4, 7, 97\}$	97	1	97	1	97	1	Yes	Yes
8) $\{(-1)^n\left(2 - \frac{1}{n}\right) : n \in \mathbb{N}\}$	2	-2	DNE	DNE	2	-2	No	Yes
9) $\{\ln(x) : x \in \mathbb{R}, x > 0\}$	DNE	DNE	DNE	DNE	DNE	DNE	✗	No
10) $\{n^{1/n} : n \in \mathbb{N}\}$	3	1	3	1	3	1	Yes	Yes
11) $\{\arctan(x) : x \in \mathbb{R}\}$	$\frac{\pi}{2}$	$\frac{-\pi}{2}$	DNE	DNE	$\frac{\pi}{2}$	$\frac{-\pi}{2}$	No	Yes
12) $\{(-1)^n : n \in \mathbb{N}\}$	1	-1	1	-1	1	-1	Yes	Yes
13) $\{e^x : x \in \mathbb{R}\}$	DNE	0	DNE	DNE	DNE	0	✗	No

$\lim \frac{\ln n}{n}$
$n \to \infty = 1$

1.3 Critical Thinking Questions

Based on your findings in Section 1.2, answer the following questions. Provide a counterexample for any of the questions for which the answer is "no." As you answer these questions, remember that our goal is to get a clear understanding of the meanings of the terms — recognizing their strengths and limitations (i.e., determining the relationship between the terms, as well as seeing all that is present but not reading more into a given definition than is present). For all the following, assume each set is a nonempty subset of $\mathbb{R}$.

1. Does a set always have a maximum? a minimum?

2. Does a set always have a supremum? an infimum?

3. If a set is bounded, must it possess a supremum/infimum?

4. If a set has a supremum (infimum), must the set be bounded above (below)?

5. If a set is bounded, must it possess a maximum and minimum?

6. If a set has a maximum (minimum), must the set be bounded from above (below)? Explain.

7. In questions 1–6, you have examined the relationship between boundedness and the existence of a maximum (minimum) and/or supremum (infimum). Consider the following two statements:

   (a) If a set of real numbers is bounded above, it has a maximum.

   (b) If a set of real numbers is bounded above, it has a supremum.

   Only one of these statements is true. Identify the one which is false and give a counterexample. You may use your findings from Section 1.2. The proof of the true statement is nontrivial. In fact, that statement is sometimes taken as an axiom of the real numbers. Explain why the statement seems reasonable.

8. If a set has a supremum (infimum), must it have a maximum (minimum)?

9. If a set has a maximum (minimum), must it also have a supremum (infimum)?

10. In questions 8–9, you have examined the relationship between the supremum (infimum) and maximum (minimum) of a set of real numbers. Consider the following two statements:

    (a) If a set has a supremum (infimum), then it has a maximum (minimum).

    (b) If a set has a maximum (minimum), then it has a supremum (infimum).

    Only one of these statements is true. Identify the one which is true and prove it. Give a counterexample to the statement that is false. You may use the information you obtained in Section 1.2.

11. Assuming a set has a maximum (minimum), is the maximum an element of the set?

12. Assuming a set has a supremum (infimum), is the supremum (infimum) always an element of the set?

13. In questions 11–12, you have examined the relationship between notions of boundedness and set membership. Consider the following two statements:

    (a) If a set has a supremum (infimum), then the supremum (infimum) is an element of the set.

    (b) If a set has a maximum (minimum), then the maximum (minimum) is an element of the set.

    Only one of these statements is true. Identify the one which is true and prove it. Give a counterexample to the statement that is false. You may refer to Section 1.2 to help you in making these determinations.

1.4   Questions for Reflection

1. Let A be a nonempty subset of the real numbers. Define the set $|A|$ to be $|A| := \{|x| : x \in A\}$.

   (a) If A is bounded above, is $|A|$ necessarily bounded above? If not, give an example. If so, by what? Explain.

   (b) If the set A is bounded, is the set $|A|$ bounded? If not, give an example. If so, by what? Explain.

(c) If the maximum of A exists, does $|A|$ have a maximum? If so, what would it be? If not, give an example.

(d) If the minimum of A exists, does $|A|$ have a minimum? If so, what would it be? If not, give an example.

(e) If the supremum and infimum of A exist, does $|A|$ have a supremum? If not, give an example. If so, what would it be? Does $|A|$ have an infimum? If not, give an example. If so, what would it be?

2. Let $A = [1, 4]$ and $B = [5, 11]$. Relate the set B to the set of all upper bounds of A. If a set is bounded above, does it have only one upper bound? If there is more than one, how many upper bounds does the set have?

3. Suppose a set A is nonempty and $\sup(A)$ and $\inf(A)$ exist. Suppose a number y satisfies the inequality $\inf(A) < y < \sup(A)$. Must y be in A? If your answer is no, give an example.

4. Suppose a set A is nonempty and $\sup(A)$ and $\inf(A)$ exist. Suppose a number y satisfies the inequality $\inf(A) < y < \sup(A)$. Must there exist an element x of A such that $y < x < \sup(A)$? Give an example if the answer is no.

5. Suppose there are two nonempty subsets of the real numbers L and R such that

- if $x \in L$ and $y \in R$, then $x < y$, and
- $L \cup R = \mathbb{R}$.

Answer the following questions.

(a) If the set L has a supremum, but no maximum, can the set R have an infimum? minimum? If your answer is no, explain. If your answer is yes, give an example.

(b) If the set L has a maximum, can the set R have an infimum? minimum? If your answer is no, explain. If your answer is yes, give an example.

(c) If the set R has an infimum, but no minimum, can the set L have a supremum? maximum? If your answer is no, explain. If your answer is yes, give an example.

(d) If the set R has a minimum, can the set L have a supremum? maximum? If your answer is no, explain. If your answer is yes, give an example.

(e) Is there a relationship between the infimum of R and the supremum of L?

6. Consider the following two collections of sets:

$$\{I_n = [0, 1/n] : n \in \mathbb{N}\}$$

and

$$\{J_n = (0, 1/n) : n \in \mathbb{N}\}.$$

(a) Compute $\bigcap_{n=1}^{\infty} I_n$.

(b) Compute $\bigcap_{n=1}^{\infty} J_n$.

(c) Can you explain the differences in the two answers using any of the terms introduced in this lab?

2

Introducing the "Epsilon Definition" of Least Upper Bound

2.1 Introduction

Let X be a nonempty set of real numbers. A real number b is an *upper bound* of X if for every $x \in X$, $x \leq b$. The *supremum*, or *least upper bound*, of X is the "smallest" of the upper bounds of X. The supremum of a set X, often denoted $\sup(X) = s$, must satisfy the following two conditions:

- If $x \in X$, then $x \leq s$.
- If b is any upper bound of X, then $s \leq b$.

In this lab, you are asked to formulate a second, equivalent definition of supremum. This definition will be useful in proving a number of theorems, as well as to help you better understand the concepts of supremum and infimum. For instance, exactly what do we mean when we say that the supremum is the "smallest" upper bound? How does one determine the supremum? How does one prove the existence of the supremum of a set? Is there a difference between whether the supremum is or is not an element of the set? The definition you write in this lab will help you to answer these questions. Moreover, this definition is similar in several respects to definitions involving the convergence of a sequence and the limit of a function, two fundamental concepts that will be considered in later labs and used throughout the course.

2.2 Using Examples to Enhance Understanding

2.2.1 Example Set 1

Let $S_1 = \left\{ \dfrac{n}{n+1} : n \in \mathbb{N} \right\}$.

5

1. According to the definition of supremum given in the Introduction, what is the supremum (or least upper bound) of S_1? We will denote the supremum of this and other sets considered in this lab by the letter s. $s = 1$

2. For the remainder of this lab, we will let the Greek letter ϵ designate a positive real number. If we let $\epsilon = .5$, can you find an element of S_1 that lies in the half-open interval $(s - \epsilon, s]$; that is, can you find elements of S_1 that are larger than $s - \epsilon$ and less than or equal to s? If so, describe all such elements of S_1 that satisfy this condition. If you cannot find any such elements, explain why. $S_1 = \left\{ \frac{n}{n+1} ; n \in \mathbb{N} \; n > 1 \right\}$

3. Repeat exercise 2 for $\epsilon = .1, .05, .01$.

4. For $s = \sup(S_1)$, does is it seem possible, based on the data you gathered in exercises 2 and 3, to find an $\epsilon > 0$ for which no elements of S_1 lie in $(s - \epsilon, s]$? If so, describe all such ϵ, and explain why there are no elements of S_1 that lie in the interval $(s - \epsilon, s]$. If the answer is no, try to explain what it is about the nature of the supremum that makes it always seem possible to find elements of the set S_1 in the interval $(s - \epsilon, s]$ for every $\epsilon > 0$.

5. Choose an upper bound u of S_1 that is *not* equal to the supremum, and repeat exercises 2, 3, and 4.

6. For any upper bound $u \neq \sup(S_1)$, does is it appear possible to find values of $\epsilon > 0$ for which no elements of S_1 lie in the half-open interval $(u - \epsilon, u]$? Explain your answer. Based on your findings, does there appear to be a difference in the behavior between the supremum and an arbitrary upper bound, at least as it relates to the issue of whether we can find elements of the set S_1 in the interval $(s - \epsilon, s]$ for any value of $\epsilon > 0$?

2.2.2 Example Set 2

1. For each set in Table 2.1, identify the supremum s, and enter your response in column 2. For each value of ϵ given in columns 3, 4, and 5, determine whether there are elements of the set that fall in the half-open interval $(s - \epsilon, s]$. If so, enter yes in the appropriate cell of the table, and then describe all such elements of the set that satisfy this condition. If there are no such elements, enter no in the table, and provide an explanation as to why this might be the case.

TABLE 2.1

Set	s	$\epsilon = .5$	$\epsilon = .1$	$\epsilon = .05$	Q2
$S_2 = \{x \in \mathbb{R} : 0 \leq x < 1\}$					
$S_3 = \left\{ (-1)^n \left(2 + \frac{1}{n}\right) : n \in \mathbb{N} \right\}$					
$S_4 = \{\arctan(x) : x \in \mathbb{R}\}$					
$S_5 = \{(-1)^n : n \in \mathbb{N}\}$					

2. For each set in Table 2.1, is it possible, based on the data you have entered, to find an $\epsilon > 0$ for which no elements of the given set lie in $(s - \epsilon, s]$? If so, enter yes in column 6 (under Q2), and describe all such ϵ. If not, enter no in column 6. In either case, explain your results.

3. For each set in Table 2.2, select an upper bound u that is *not* equal to the supremum. Enter this value in column 2. For each ϵ given in columns 3, 4, and 5, determine whether there are elements of the set that fall in the half-open interval $(u - \epsilon, u]$. If so, enter yes in the appropriate cell of the table, and then describe all such elements of the set that satisfy this condition. If there are no such elements, enter no in the table, and provide an explanation as to why this might be the case.

TABLE 2.2

Set	u	$\epsilon = .5$	$\epsilon = .1$	$\epsilon = .05$	Q4
$S_2 = \{x \in \mathbb{R} : 0 \leq x < 1\}$					
$S_3 = \{(-1)^n (2 + \frac{1}{n}) : n \in \mathbb{N}\}$					
$S_4 = \{\arctan(x) : x \in \mathbb{R}\}$					
$S_5 = \{(-1)^n : n \in \mathbb{N}\}$					

4. For any upper bound u that is *not* the supremum, does it seem possible, based on the data in Table 2.2, to find an $\epsilon > 0$ for which no elements of the given set lie in $(u - \epsilon, u]$? If so, enter yes in column 6 (under Q4), and describe all such ϵ. If not, enter no in column 6. In either case, explain your results.

5. Compare and contrast your findings for the supremum and an arbitrarily chosen upper bound. In particular, does there appear to be a difference in the behavior between the supremum and an arbitrary upper bound, at least as it relates to the issue of whether we can find elements of the set S_i ($i = 2, 3, 4, 5$) in the interval $(s - \epsilon, s]$ for any value of $\epsilon > 0$?

2.3 Critical Thinking Questions

Based on the data you gathered in the Section 2.2, answer the following questions.

1. If s is the least upper bound of a nonempty set X of real numbers, and if $\epsilon > 0$, can we always find elements of X in the half-open interval $(s - \epsilon, s]$? Why, or why not?

2. If u is an arbitrarily chosen upper bound of X that is not equal to the supremum, and if $\epsilon > 0$, can we always find elements of X in the half-open interval $(u - \epsilon, u]$? Why, or why not?

3. If your answers to questions 1 and 2 differ, can you explain why? If what way does the supremum differ from any other upper bound?

4. Writing the definition: Based on your answers to the prior three questions, try to write the "new" definition. The statement of this definition will involve a nonempty set X, a positive real number ϵ, and the half-open interval $(s - \epsilon, s]$, where s denotes the supremum.

2.4 Equivalent Definitions

Two statements p and q are defined to be equivalent if the *biconditional* statement $p \Longleftrightarrow q$ (p if and only if q) is true. The biconditional $p \overset{\scriptscriptstyle i}{\Longleftrightarrow} q$ is shorthand for the conjunction: If p then q, *and* if q then p. In order to show that two definitions D_1 and D_2 are equivalent, we must prove that the biconditional $D_1 \Longleftrightarrow D_2$ is true. Let us call the definition given in the Introduction Definition 1 and the one you just formulated Definition 2.

Definition 1: Let X be a nonempty set of real numbers. The number s is called the supremum of X if s is an upper bound of X and $s \le y$ for every upper bound y of X.

Definition 2: This is the "new" definition that you derived in Section 2.3.

In order to show that the two definitions are equivalent, we must prove the following two conditional statements:

(i) If $s = \sup(X)$, as given by Definition 1, then s is the supremum, as given by Definition 2. Here, *assume that Definition 1 holds*, and use this assumption *to prove that Definition 2 holds*. (Definition 1 $\Longrightarrow$ Definition 2)

(ii) If $s = \sup(X)$, as given by Definition 2, then s is the supremum, as given by Definition 1. Here, *assume that Definition 2 holds*, and use this assumption *to prove that Definition 1 holds*. (Definition 2 $\Longrightarrow$ Definition 1)

What is the practical significance of showing that these two definitions are logically equivalent?

2.5 Questions for Reflection

1. If X is a nonempty set for which the supremum exists, that is, $\sup(X) = s$, what is the minimal number of elements of X, if any, that must lie in $(s - \epsilon, s]$ for every choice of $\epsilon > 0$? Carefully explain your response. Is your answer consistent with the definition you wrote in Section 2.3?

2. Use the "new" definition to prove that

$$\sup\left(\left\{\left(1 - \frac{1}{3^n}\right) : n \in \mathbb{N}\right\}\right) = 1.$$

3. Identify the supremum of the set given below, and use the "new" definition of supremum to prove your claim.

$$\left\{\left(-\frac{1}{2}\right)^n : n \in \mathbb{N}\right\}$$

4. In considering the "new" definition of supremum, state a condition, or set of conditions, by which a set would fail to have a supremum? In other words, what is the negation of the definition you came up with in Section 2.3?

5. If s is the supremum of a nonempty set X of real numbers, under what condition(s) do we find an infinite number of elements of X in the interval $(s - \epsilon, s]$ for every $\epsilon > 0$? Justify your reasoning.

6. If X is the empty set, does the supremum of X exist? Explain your answer.

7. Formulate the two definitions of infimum (greatest lower bound), and prove that they are equivalent.

3

Introduction to the Formal Definition of Convergence

3.1 Introduction

Consider the sequence $\left(\frac{1}{n}\right)_{n=1}^{\infty}$. To list some of its terms, we apply the formula $\frac{1}{n}$ to successive positive integer values n. For instance, if we set $n = 1$, the formula yields 1; if we set $n = 2$, the formula yields $\frac{1}{2}$. If we continue, we get the list:

$$1, \frac{1}{2}, \frac{1}{3}, \frac{1}{4}, \ldots$$

The relationship between each positive integer n and the corresponding term $\frac{1}{n}$ suggests a functional relationship. This is in fact the case, as a sequence is a function whose domain is the set of positive integers. The *closed form notation* $(a_n)_{n=1}^{\infty}$ reveals these components: to each integer input $n = k$, the term a_k is assigned as output. Here, the ∞ symbol indicates that the assignment process, $1 \mapsto a_1, \ 2 \mapsto a_2, \ldots, \ k \mapsto a_k, \ldots$, continues indefinitely.

A sequence, like any single-variable function, can be graphed in the xy-coordinate plane. The first 10 points of the graph of the sequence $\left(\frac{1}{n}\right)_{n=1}^{\infty}$ are displayed below.

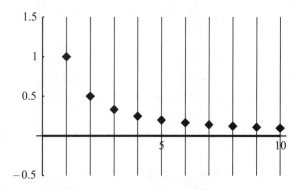

The portion of the graph shown above suggests that the terms of the sequence $\left(\frac{1}{n}\right)_{n=1}^{\infty}$ tend toward zero as n increases. This is consistent with the idea of convergence you encountered in calculus: a sequence $(a_n)_{n=1}^{\infty}$ *converges* to some real number L, provided that the terms a_n get "closer and closer" to L as n "increases without bound." Symbolically, this is represented using the notation $\lim_{n\to\infty} a_n = L$, where L denotes the *limit of the sequence*. If there is no such finite number L to which the terms of the sequence get arbitrarily close, then the sequence is said to *diverge*.

The problem with this characterization is its imprecision. Exactly what does it mean for the terms of a sequence to get "closer and closer," or "as close as we like," or "arbitrarily close" to some finite number L? Is there a point beyond which every term is equal to L, or is it enough for the terms to get within a certain, prescribed distance? Must every term get close to L, or is it possible for some, or even many, of the terms to approach some other number, or maybe even no particular number at all? Even if we accept this apparent ambiguity, how would one use the definition given in the preceding paragraph to prove theorems that involve sequences?

Since sequences are used throughout analysis, the concepts of convergence and divergence must be carefully defined. The purpose of this lab is to investigate how to make these notions precise.

Note: For some of the activities in this lab, you may wish to use the accompanying *Visual Guide for Lab 3*, which can be found at www.saintmarys.edu/~jsnow. The *Visual Guide* also appears in the appendix. Alternatively, you may use the *Maplet for Lab 3*, which can be found at www.saintmarys.edu/~jsnow.

3.2   Using Examples to Enhance Understanding

3.2.1   Example Set 1

Consider the following two sequences.

Sequence 1:
$$\left(\frac{(-1)^n}{n}\right)_{n=1}^{\infty}$$

Sequence 2:
$$\left((-1)^n\left(2+\frac{1}{n}\right)\right)_{n=1}^{\infty}$$

Sequence 1 converges. Use your calculus knowledge to identify the limit. Sequence 2 diverges. Before going on, think about why this is the case. Use this information to complete the following series of questions. Enter your answers in the table that follows. Here, ϵ refers to a positive real number.

1. Let $\epsilon = .5$. For Sequence 1, the limit of the sequence is $L = 0$. Your *Visual Guide* gives code for graphing Sequence 1. In addition to a graph of the sequence, you will want to enter the code necessary to graph the horizontal lines $y = L - .5$, $y = L$, and $y = L + .5$. A completed graph will look like the one presented in the following figure. For the sake of simplicity, we refer to the horizontal region denoted by $(L - .5, L + .5)$ as an "ϵ-band." Your task for this exercise is to identify those values of n whose corresponding terms a_n fall within the interval $(L - .5, L + .5)$. Record your answers in the appropriate cell of the column labeled $\epsilon = .5$.

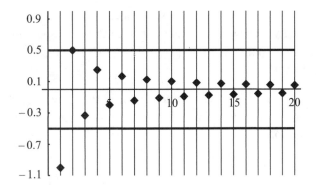

2. Although Sequence 2 diverges, repeat question 1 using $L = 2$. As you did with Sequence 1, make sure that your graph includes the horizontal lines $y = L - .5$, $y = L$, and $y = L + .5$. For the ϵ-band, which in this case is $(2 - .5, 2 + .5)$, identify those values of n whose corresponding terms a_n fall within the given interval $(2 - .5, 2 + .5)$, and record your answers in the appropriate cell of the column labeled $\epsilon = .5$.

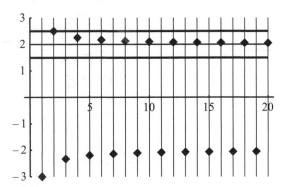

3. Repeat question 1 for Sequence 1, with $L = 0$ and $\epsilon = .2, .1, .05$. In each case, identify those values of n whose corresponding terms a_n fall within the prescribed ϵ-band, and record your answers in the appropriate cells in columns 3–5.

4. Repeat question 2 for Sequence 2, with $L = 2$ and $\epsilon = .2, .1, .05$. In each case, identify those values of n whose corresponding terms a_n fall within the prescribed ϵ-band, and record your answers in the appropriate cells in columns 3–5.

Sequence	Identification of those n for which $a_n \in (L - \epsilon, L + \epsilon)$				Q5	Q6
	$\epsilon = .5$	$\epsilon = .2$	$\epsilon = .1$	$\epsilon = .05$		
Sequence 1 $L = 0$	$n > 2$	$n > 5$	$n > 10$	$n > 20$	Yes	No
Sequence 2 $L = 2$	n is even $n > 2$	n is even $n > 5$	n is even $n > 10$	n is even $n > 20$	No	Yes

5. Consider the data you obtained for each sequence. For each ϵ considered in the table, is there a value of n beyond which *all* a_n are contained in the corresponding ϵ-band? In other words, for all n beyond some point, is $a_n \in (L - \epsilon, L + \epsilon)$? Enter your answer, either yes or no, in the column labeled Q5.

6. For either sequence, and using the ϵ values given in the table, would your answers to question 5 change if L had been assigned a different value? Enter your answer, either yes or no, in the column labeled Q6.

3.2.2 Example Set 2

Consider the following four sequences, and answer the accompanying questions.

Sequence 3:
$$\left(\frac{n}{n+1}\right)_{n=1}^{\infty} \quad \to 1$$

Sequence 4:
$$\left(\frac{\sin\left(\frac{n\pi}{2}\right)}{n}\right)_{n=1}^{\infty} \quad \to 0$$

Sequence 5:
$$\left(\frac{n!}{25^n}\right)_{n=1}^{\infty}$$

Sequence 6:
$$\left(\begin{cases} 1, & \text{if } n \text{ is odd} \\ \dfrac{1}{n}, & \text{if } n \text{ is even} \end{cases}\right)_{n=1}^{\infty}$$

Both Sequences 3 and 4 converge. Use your calculus knowledge to identify the limit of each sequence. Sequences 5 and 6 both diverge. Before going on, try to reflect on why this is the case.

1. For each sequence, each value of ϵ, and each value of L specified in the table that follows, graph the sequence, the ϵ-band $(L - \epsilon, L + \epsilon)$, and L. For Sequences 3 and 4, L is the limit of convergence. For Sequences 5 and 6, L is not the limit; it simply denotes a value about which you are being asked to construct each ϵ-band. In each case, identify those values of n for which the terms a_n fall within the region $(L - \epsilon, L + \epsilon)$. Record your answers in columns 2–4 of the table below.

2. Consider the data you have recorded for each sequence. For each ϵ given in the table, is there a value of n beyond which *all* a_n are contained in the corresponding ϵ-band? In other words, for all n beyond some point, is $a_n \in (L - \epsilon, L + \epsilon)$? Enter your answer, either yes or no, in the column labeled Q2.

3. Consider each sequence. Using the ϵ values given in the table, would your answers to question 2 change if L had been assigned a different value? Enter your answers, either yes or no, in the column labeled Q3.

Sequence	Identification of those n for which $a_n \in (L - \epsilon, L + \epsilon)$			Q2	Q3
	$\epsilon = .2$	$\epsilon = .1$	$\epsilon = .05$		
Sequence 3 $L = 1$					
Sequence 4 $L = 0$					
Sequence 5 $L = 0$					
Sequence 6 $L = 0$					

3.2.3 Example Set 3

The table given below reports information on four unknown sequences, which are referred to as Sequences 7, 8, 9, and 10. For each sequence, all that is known is the behavior of the terms with respect to some given value L (which may or may not be a limit) and the values of ϵ specified across the top. For Sequence 7, the first entry labeled "All n" means that every term of the unknown sequence falls within the ϵ-band $(2 - 2, 2 + 2)$. The next entry for Sequence 7 labeled "$n \geq 5$" (corresponding to $\epsilon = 1$) reveals that all terms beyond and including the fifth term of Sequence 7 are contained in the region $(2 - 1, 2 + 1)$. In general, every entry describes those terms contained in the region $(L - \epsilon, L + \epsilon)$. Use the data in the table to answer the accompanying questions for each of the "mystery" sequences.

Sequence	Description of those n for which $a_n \in (L - \epsilon, L + \epsilon)$					
	$\epsilon = 2$	$\epsilon = 1$	$\epsilon = .5$	$\epsilon = .1$	$\epsilon = .01$	$\epsilon = .001$
Sequence 7 $L = 2$	All n	$n \geq 5$	$n \geq 30$	$n \geq 100$	$n \geq 600$	$n \geq 1500$
Sequence 8 $L = 1$	$1 \leq n \leq 5$	$n = 1, 2$	$n = 1$	No n	No n	No n
Sequence 9 $L = 0$	All n	All n	Even n	Even n	Even n	Even n
Sequence 10 $L = -1$	$n \leq 30$	$n \leq 30$	$n \leq 30$	$n \leq 28$	$n \leq 20$	$n \leq 15$

1. Based upon the data in the table, does Sequence 7 seem to converge to $L = 2$? Why, or why not?

2. Based upon the data in the table, does Sequence 8 seem to converge to $L = 1$? Why, or why not?

3. Based upon the data in the table, does Sequence 9 seem to converge to $L = 0$? Why, or why not?

4. Based upon the data in the table, does Sequence 10 seem to converge to $L = -1$? Why, or why not?

3.3   Critical Thinking Questions

3.3.1   Formulating the Definition

1. If a sequence $(a_n)_{n=1}^{\infty}$ converges to a real number L, and if ϵ is any positive real number, what can we say about the terms of the sequence in relation to the ϵ-band $(L - \epsilon, L + \epsilon)$? Specifically, for each $\epsilon > 0$, is there a value of n such that $a_k \in (L - \epsilon, L + \epsilon)$ for all $k > n$? Carefully justify your answer using your findings from Section 3.2.

2. If a sequence $(a_n)_{n=1}^{\infty}$ diverges, and if ϵ is any positive real number, what can we say about the terms of the sequence in relation to the ϵ-band $(L - \epsilon, L + \epsilon)$ for some number L that is not the limit? In particular, does there exist an $\epsilon > 0$ for which no n can be found such that $a_k \in (L - \epsilon, L + \epsilon)$ for all $k > n$? Carefully justify your answer using your findings from Section 3.2.

3. Describe the relationship between ϵ and n. Does it appear as though ϵ is a function of n, or n is a function of ϵ?

4. Based upon your analysis of the examples in Section 3.2 and your answers to the three questions posed here, try to formulate the definition of the convergence of a sequence. A correct definition will link the limit of convergence L of a sequence $(a_n)_{n=1}^{\infty}$ to the relationship between $\epsilon > 0$ and "input" values $n \in \mathbb{N}$.

3.3.2   Comparison Between the Definition and Intuition

As mentioned in the introduction, a sequence $(a_n)_{n=1}^{\infty}$ converges to L if the terms of the sequence get "closer and closer" to L as n gets "larger and larger." In graphical terms, this means that as we move along the x-axis and away from the origin (n values increase), the resulting outputs (terms of the sequence) tend toward L. Is the definition you just formulated consistent with this notion?

To answer this question, let's consider what happens to the following sequence, which converges to 2.

Sequence:
$$\left(2 - \frac{1}{n}\right)_{n=1}^{\infty}$$

Let $\epsilon = .2$. Find the smallest n_1 such that $a_n \in (2 - .2, 2 + .2)$ for $n \geq n_1$?

Let $\epsilon = .1$. Find the smallest n_2 such that $a_n \in (2 - .1, 2 + .1)$ for $n \geq n_2$?

Let $\epsilon = .05$. Find the smallest n_3 such that $a_n \in (2 - .05, 2 + .05)$ for $n \geq n_3$?

Let $\epsilon = .01$. Find the smallest n_4 such that $a_n \in (2 - .01, 2 + .01)$ for $n \geq n_4$?

Use the information regarding n_1, n_2, n_3, and n_4 to fill in the blanks below.

Terms a_1 to _____ are .2 or more units from $L = 2$.

Terms _____ to _____ are less than or equal to .2 units from $L = 2$, but are also .1 or more units away from $L = 2$.

Terms _____ to _____ are less than or equal to .1 units from $L = 2$, but are also .05 or more units from $L = 2$.

Terms _____ to _____ are less than or equal to .05 units from $L = 2$, but are also .01 or more units from $L = 2$.

Terms _____ to _____ are less than or equal to .01 units from $L = 2$.

Is your ability to select a corresponding value of n for each given value of ϵ consistent with the definition you wrote in Section 3.3? If so, does the procedure obtained by selecting successively smaller values of ϵ yield results consistent with the notion of convergence described in the Introduction? Be sure to justify your responses.

3.4 Questions for Reflection

1. Suppose ϵ could only take on the values .2, .1, .01, .001, .0001. If it were possible to write a computer program that uses the definition you wrote in Section 3.3 to check the convergence of a sequence, what would the steps of the program look like?

2. For each of the following statements, identify the error, and explain how the definition is violated.

   (a) Let $(a_n)_{n=1}^{\infty}$ be a sequence, and suppose that $L = 2$. For every $\epsilon > 0$, there exists an N such that $a_n \in (2 - \epsilon, 2 + \epsilon)$ for infinitely many $n > N$. Therefore, $(a_n)_{n=1}^{\infty}$ converges to 2.

   (b) Let $(b_n)_{n=1}^{\infty}$ be a sequence, and suppose that $L = 3$. For every $\epsilon > .5$, there exists an N such that $b_n \in (3 - \epsilon, 3 + \epsilon)$ for all $n > N$. Therefore, $(b_n)_{n=1}^{\infty}$ converges to 3.

   (c) Let $(c_n)_{n=1}^{\infty}$ be a sequence, and suppose that $L = -2$. For every N, we can find $\epsilon > 0$ such that $c_n \in (-2 - \epsilon, -2 + \epsilon)$ for all $n > N$. Therefore, $(c_n)_{n=1}^{\infty}$ converges to -2.

   (d) Let $(d_n)_{n=1}^{\infty}$ be a sequence, and suppose that $L = 5$. There is an $\epsilon > 0$ such that $d_n \in (5 - \epsilon, 5 + \epsilon)$ for all $n > N$. Therefore, $(d_n)_{n=1}^{\infty}$ converges to 5.

3. For each part in question 2, give an example of a divergent sequence that satisfies the given conditions.

4. Use the definition you formulated in Section 3.3 to prove that the sequence

$$\left(\frac{1}{3^n} \right)_{n=1}^{\infty}$$

converges to 0.

4

Experience with the Definition of the Limit of a Sequence

4.1 Introduction

The notion of the limit is the highlight of calculus. To understand the limit is to understand the crux of calculus. In Lab 3, we started on that path of understanding by considering the special case of the limit of a sequence. While a sequence can have a limit of ∞ or $-\infty$, in this lab, as in Lab 3, we consider only sequences with a finite limit; that is, the limit is a real number. When the limit of a sequence is a real number, we say the sequence is convergent.

The definition of the limit of a sequence is complicated because it involves three quantifiers: two universal quantifiers and one existential quantifier. If we understand the roles of the quantifiers, we have a handle on the definition with regard to what it means and how to apply it. The purpose of this lab is twofold: to increase your understanding of the definition of the limit of a sequence by focusing on the relationship between ϵ and N and to develop an algorithm to prove that a given number is the limit of a given sequence. We examine the graphs of several sequences to help us reach these goals before proceeding to an algebraic approach.

This lab may be treated as a continuation of the prior lab, in which the purpose was to formulate the definition of the limit of a sequence. This lab is devoted to learning how to use the definition to construct proofs.

We recall the definition of the limit of a sequence of real numbers.

> Let $(s_n)_{n=1}^{\infty}$ be a sequence of real numbers. We say that the sequence *converges to L*, or the real number L is the *limit of the sequence*, if for every $\epsilon > 0$, there exists a number N such that if $n > N$, then $|s_n - L| < \epsilon$. We write $\lim_{n \to \infty} s_n = L$.

For this lab, you may wish to use the accompanying *Visual Guide for Lab 4*, which can be found at www.saintmarys.edu/~jsnow. The *Visual Guide* for this section also appears in the appendix. Alternatively, you may use the *Maplet for Lab 4*, which can be found at www.saintmarys.edu/~jsnow.

4.2   Using Examples to Enhance Understanding

In this part of the lab, you look at the steps that one would follow in applying the definition graphically. First, you use a graph to determine the limit of each sequence. Then, for each specified $\epsilon > 0$, you are asked to find a corresponding N value, as prescribed by the definition. This is similar to what you did in Lab 3. The difference here is that you are asked to justify your choice of N. This is an important step in learning how to apply the definition algebraically. However, you are still not proving that the limit is the claimed value, since you are only considering a finite number of epsilon values. Our goal in this part is simply to understand how the definition works. For help with the code, see the accompanying Visual Guide or the Maplet for Lab 4. If you use the Visual Guide, save the computer worksheet you generate from this lab on your personal computer space. It may be helpful in later labs.

Consider the following sequences:

i) $(a_n)_{n=1}^{\infty} = \left(\dfrac{1}{n} \right)_{n=1}^{\infty}$

    ii) $(b_n)_{n=1}^{\infty} = \left(\dfrac{5n-2}{n+4} \right)_{n=1}^{\infty}$

iii) $(c_n)_{n=1}^{\infty} = \left(\dfrac{(-1)^n}{\sqrt{n}} \right)_{n=1}^{\infty}$

    iv) $(d_n)_{n=1}^{\infty} = \left(\begin{cases} 3 - n^{-2/3}, & \text{if } n \text{ is even} \\ 3, & \text{if } n \text{ is odd} \end{cases} \right)_{n=1}^{\infty}$

v) $(e_n)_{n=1}^{\infty} = \left(4^{1/n} \right)_{n=1}^{\infty}$

    vi) $(f_n)_{n=1}^{\infty} = \left(\dfrac{100^n}{n!} \right)_{n=1}^{\infty}$

1. For each of the sequences listed above, complete steps (a)–(d). Record your answers in the table at the end of the problem. In the following questions, we use the notation (s_n) to represent the sequence being considered.

    (a) Graph the sequence for many values of n; generally, $n = 1$ to 100 is sufficient. Looking at the graph, determine the limit. Call it L. Record your result in column 2 in the table below.

    (b) Let $\epsilon = .2$. Using the definition, find an N such that if $n > N$, then $|s_n - L| < \epsilon$. You will do this graphically by plotting on the same graph the sequence and the lines: $y = L+\epsilon$ and $y = L-\epsilon$. From this graph, you should be able to determine a value for N. (Note: You may need to adjust the domain range. For example, you may need to consider small values of n such as $n \in [1, 25]$ or large values of n such as $n \in [100, 150]$.) Once you have found a value for N, record it in column 3 in the table.

    (c) Repeat Part (b) for $\epsilon = .1$ and $.05$.

    (d) For the N value that you determined for $\epsilon = .05$, choose an integer greater than N. We will call this integer "test n." For that test n, compute $|s_n - L|$. Is this number less than $.05$ as it should be? Record your answers in columns 6–8 of the table.

2. Choose a second "test n" different than the one you just selected. Do you again find that $|s_n - L| < .05$? Is it possible to find any integer greater than N, call it m, such that $|s_m - L| \geq .05$? Why or why not?

Sequence	Limit or L	N value for $\epsilon = .2$	N value for $\epsilon = .1$	N value for $\epsilon = .05$	Test n	$\|s_n - L\|$ for test n	Is $\|s_n - L\| < .05$?
$(a_n)_{n=1}^{\infty}$							
$(b_n)_{n=1}^{\infty}$							
$(c_n)_{n=1}^{\infty}$							
$(d_n)_{n=1}^{\infty}$							
$(e_n)_{n=1}^{\infty}$							
$(f_n)_{n=1}^{\infty}$							

3. In the exercises above, you did a visual check to determine N. Then you checked two specific "test n" greater than N. Why is this not sufficient to prove that the N you chose satisfies the definition?

4. We have applied the definition of limit to some specific values for epsilon. If $(s_n)_{n=1}^{\infty}$ is a sequence of real numbers for which we believe $\lim_{n \to \infty} s_n = L$, list the steps that would be employed to verify that L is the limit.

4.3　An Algebraic Way to Find N

In Section 4.2, we looked at graphs to find a value of N corresponding to a specific value of epsilon. In this section we examine an algebraic method for finding N. This algebraic work is the "scratch work" needed to write a formal proof. We follow the scratch work with a formal proof. We use two examples to explain the process.

Model A.　Consider the sequence $\left(\dfrac{10n - 3}{n} \right)_{n=1}^{\infty}$.

Scratch work to find N: We think the limit of the sequence $\left(\frac{10n-3}{n} \right)_{n=1}^{\infty}$ is 10. Let us choose a generic $\epsilon > 0$. Our goal is to find N such that

$$\text{if } \quad n > N, \quad \text{then} \quad \left| \frac{10n - 3}{n} - 10 \right| < \epsilon. \tag{1}$$

Notice that the only unknown in inequality (1) is n. So it seems logical to simply solve for n. Let us do that:

$$\left| \frac{10n - 3}{n} - 10 \right| < \epsilon \quad \Longleftrightarrow$$

$$\left| \frac{10n - 3 - 10n}{n} \right| < \epsilon \quad \Longleftrightarrow$$

$$\left| \frac{-3}{n} \right| < \epsilon.$$

Because $n > 0$, the last inequality yields the following:

$$\frac{3}{n} < \epsilon \quad \Longleftrightarrow$$

$$\frac{n}{3} > \frac{1}{\epsilon} \quad \Longleftrightarrow$$

$$n > \frac{3}{\epsilon}.$$

So we may choose N to be any number greater than or equal to $\frac{3}{\epsilon}$.

Formal Proof.   In writing this proof, we are merely verifying the limit definition for a particular sequence. (You should convince yourself of this fact by comparing the proof and the definition.)

Prove: $\lim\limits_{n \to \infty} \dfrac{10n - 3}{n} = 10.$

Proof. Let $\epsilon > 0$. Let $N = \frac{3}{\epsilon}$. Choose $n \in \mathbb{N}$ such that $n > N$. Then

$$\left| \frac{10n - 3}{n} - 10 \right| = \left| \frac{10n - 3 - 10n}{n} \right| = \left| \frac{-3}{n} \right| = \frac{3}{n} < \frac{3}{N} = \frac{3}{3/\epsilon} = \epsilon.$$

Q.E.D.

Model B.   Sometimes it is hard or messy to solve for n. For example, consider the sequence

$$\left(\frac{10n - 3}{4n + 7} \right)_{n=1}^{\infty}.$$

Scratch work to find N: We think the limit of the sequence is $5/2$. Let us choose a generic $\epsilon > 0$. As before, our goal is to find N such that

$$\text{if} \quad n > N, \quad \text{then} \quad \left| \frac{10n - 3}{4n + 7} - \frac{5}{2} \right| < \epsilon.$$

Cleaning up this inequality, we obtain

$$\left| \frac{-41}{8n + 14} \right| < \epsilon. \tag{2}$$

It would be messy to solve for n in this inequality. We find a "neat" upper bound for $\left| \frac{-41}{8n+14} \right|$ with which it is easier to work. To bound a fractional expression, we recall that a fraction is increased either when the denominator is decreased or the numerator is increased or both operations are performed. In this case, decreasing the denominator by 14 produces the "neat" upper bound: $\frac{41}{8n}$. If $\frac{41}{8n} < \epsilon$, then certainly the smaller value $\left| \frac{-41}{8n+14} \right|$ will be less than epsilon. We can easily solve for n in the inequality

$$\frac{41}{8n} < \epsilon. \tag{3}$$

We find

$$n > \frac{41}{8\epsilon}.$$ (4)

So, we choose $N = \frac{41}{8\epsilon}$.

Questions:

1. Fill in the steps from inequality (2) to inequality (3) in the scratch work above.

2. Write the formal proof for Model B.

3. Consider the expression:

$$\left| \frac{-15}{3n - 7} \right|.$$

Find a "neat" upper bound for the expression. (Hint: Decrease the denominator. You are allowed to alter coefficients.)

4. Consider the expression:

$$\left| \frac{4n + 3}{8n^2 - 7n} \right|.$$

Find a "neat" upper bound for the expression. (Hint: You will need to alter both the numerator and the denominator. Also remember the bound need only apply for n sufficiently large.)

4.4 Critical Thinking Questions

1. For $\epsilon > 0$, try to find N algebraically for each of the sequences (i)–(v) given in Section 4.2. That is, carry out the scratch work as we did in the examples in Section 4.3. Write your answers in column 2 in the table that follows. You will complete column 3 in question 3.

Sequence	Formula for N value	N for $\epsilon = .05$
$(a_n)_{n=1}^{\infty}$		
$(b_n)_{n=1}^{\infty}$		
$(c_n)_{n=1}^{\infty}$		
$(d_n)_{n=1}^{\infty}$		
$(e_n)_{n=1}^{\infty}$		
$(f_n)_{n=1}^{\infty}$		

2. Write a formal proof of the limit claim for the sequences (i)–(v) of Section 4.2.

3. Now we prepare to compare our graphical work in Section 4.2 with our algebraic work in this section. In question 1 in this section, you found a "formula" for N that depends on the choice of ϵ. So if you were given a specific ϵ, you should be able to substitute that value into the formula and generate a specific N value. For each sequence, use the formula you have entered in column 2 in the table above to determine a specific value of N corresponding to $\epsilon = .05$. Enter this specific value in column 3. Does the value of N produced by the formula result in $|s_n - L| < .05$ for all $n > N$?

4.5 Questions for Reflection

1. There may be a difference between the value you obtained for N in question 3 of Section 4.4 and the value you obtained for N through your numerical calculations in question 1 in Section 4.2. Explain why this is not a contradiction.

2. Assume $\lim_{n \to \infty} s_n = L$. For a given $\epsilon > 0$, for how many N can the statement $|s_n - L| \geq \epsilon$ for all $n > N$ be true?

3. Throughout this lab, we have acted as if the limit of a sequence is unique. Prove that the limit of a sequence is indeed unique.

4. Can a sequence attain its limit; that is, can the limit of a sequence be one of the terms of the sequence? If so, give an example. If not, explain.

5. If a sequence never attains its limit, could its terms consist of a finite number of distinct values? If so, give an example. If not, explain.

6. If a sequence (s_n) has a limit, call it L, do we have any control over the values the sequence can assume? That is, can we bound the sequence terms in any way? For example, if $\epsilon = 1$, for how many terms of the sequence is it true that $|s_n| < |L| + 1$? Is it possible to "neatly" describe those n for which the inequality $|s_n| < |L| + 1$ is true? For how many terms of the sequence does the inequality $|s_n| < |L| + 1$ fail? Based upon your answers to these questions, is the statement—a convergent sequence is bounded —true or false? If the answer is true, provide a proof. If the answer is false, identify a convergent sequence that is not bounded.

5

Experience with the Negation of the Definition of Convergence

5.1 Introduction

Examining the negation of a definition often provides insight into the definition. Showing that a definition fails to hold requires that one has a good understanding of the essential features of the terms used in the definition, as well as its logical structure. Moreover, in a proof by contradiction, one may need to use the negation. In this lab, we focus on the negation of the definition of convergence to prove that a divergent sequence does not converge to a particular value. Later in the lab, you are asked to generalize this to show that a divergent sequence does not approach any value. As a byproduct, you should better understand the logical structure of the definition itself. Recall that a sequence $(s_n)_{n=1}^{\infty}$ of real numbers is said to *converge to a real number* L provided that for every $\epsilon > 0$, there exists a number N such that $n > N$ implies $|s_n - L| < \epsilon$. In this case we write $\lim_{n \to \infty} s_n = L$, and call L the *limit* of the sequence.

For this lab, you may wish to use the accompanying *Visual Guide for Lab 5*, which can be found at www.saintmarys.edu/~jsnow. The *Visual Guide* for this section also appears in the appendix. Alternatively, you may use the *Maplet for Lab 5,* which can be found at www.saintmarys.edu/~jsnow.

5.2 Using Examples to Enhance Understanding

First we examine the situation where someone has proposed an incorrect value for the limit of a sequence. The goal is to show that person how their proposed value violates the definition.

1. Consider the sequence $(a_n)_{n=1}^{\infty} = (1/n)_{n=1}^{\infty}$. Suppose someone proposed that $L = .3$, i.e., $\lim_{n \to \infty} a_n = .3$. We wish to show the person the error of this statement.

   (a) Let $\epsilon = .5$. Now let $N = 5$. Is it true that for all $n > N$, we have $|a_n - L| < .5$? If the answer is no, give an n for which the inequality ($|a_n - L| < .5$) fails to hold. Answer the same question for $N = 10$, $N = 50$, $N = 100, N = 500$.

(b) Repeat the directions in part (a) for $\epsilon = .4$.

(c) Repeat the directions in part (a) for $\epsilon = .3$.

(d) Repeat the directions in part (a) for $\epsilon = .1$.

2. Consider the sequence

$$(b_n)_{n=1}^{\infty} = \left(\frac{2n+1}{3n-2}\right)_{n=1}^{\infty}.$$

Suppose someone proposed that $L = 1$; i.e., $\lim_{n\to\infty} b_n = 1$.

(a) Let $\epsilon = .5$. Now let $N = 5$. Is it true that for all $n > N$, we have $|b_n - L| < .5$? If the answer is no, give an n for which the inequality ($|b_n - L| < .5$) fails to hold. Answer the same question for $N = 10$, $N = 50$, $N = 100$, $N = 500$.

(b) Repeat the directions in part (a) for $\epsilon = .4$.

(c) Repeat the directions in part (a) for $\epsilon = .3$.

(d) Repeat the directions in part (a) for $\epsilon = .1$.

3. Repeat the directions in question 1 for the sequence

$$(c_n)_{n=1}^{\infty} = \left(\frac{6}{\sqrt{n}}\right)_{n=1}^{\infty}$$

with the supposed limit being .2.

5.3   Critical Thinking Questions

In this section, we examine the numerical results from Section 5.2. Our goal is to use this information to help formulate the negation of the definition of convergence.

1. Looking at the results from Section 5.2, answer the following questions and record your answers in the table which follows.

   (a) For the sequence $(a_n)_{n=1}^{\infty}$ and $\epsilon = .5$, did you find any N for which $n > N$ implied $|a_n - L| < .5$? If it was true for some N, record the smallest such N in column 2 of the table. Otherwise, write "no N worked." Answer the same question for $\epsilon = .4, .3, .1$ and record your responses in columns 3–5 of the table.

   (b) Repeat part (a) for sequences $(b_n)_{n=1}^{\infty}$ and $(c_n)_{n=1}^{\infty}$ and record your answers in the appropriate cells in the table.

2. Looking at your answers in the table above, answer the following questions regarding use of the phrase "no N worked."

   (a) For each sequence, was every ϵ flawed in the sense that it was not possible to find an N to make the condition "if $n > N$, then $|a_n - L| < \epsilon$" true?

   (b) Sometimes you wrote in the table that "no N worked." Would you have obtained the same answers if you had used larger values of N?

   (c) In those cases in which you wrote "no N worked," how did you reach this conclusion?

3. Suppose $(s_n)_{n=1}^{\infty}$ is a sequence whose terms do not approach a given real number L; that is, if the limit of the sequence $(s_n)_{n=1}^{\infty}$ exists, $\lim_{n\to\infty} s_n \neq L$. Answer the following questions in relation to $(s_n)_{n=1}^{\infty}$ and L.

(a) How many values of epsilon do we need to consider in order to show that L is not the limit of $(s_n)_{n=1}^{\infty}$? Do we need to consider a specific epsilon, finitely many epsilons, some infinite family of epsilons, or all $\epsilon > 0$?

(b) If the statement of the definition fails for a particular value of ϵ, say ϵ_*, for how many values of N is the conditional statement — If $n > N$, then $|s_n - L| < \epsilon_*$ — false? Is the statement false for every N, one value for N, infinitely many N, some finite set of N?

(c) To convince yourself that the conditional statement — If $n > N$, then $|s_n - L| < \epsilon_*$ — is false for a given N, how many $n > N$ must you consider? Must you consider all $n > N$, one n such that $n > N$, or infinitely many $n > N$?

4. Using your responses to questions 1–3 above, write the negation of the statement: the sequence $(s_n)_{n=1}^{\infty}$ converges to the real number L. Pay careful attention to the quantifiers.

5. Using the negation of the definition of convergence, prove each of the statements (a)–(c). In each case, note that if the sequence has a limit, it is not the value suggested. You may find that graphing the sequence and experimenting with some epsilon bands may help you with the proof.

(a) $\lim\limits_{n\to\infty} (1/n) \neq .3$.

(b) $\lim\limits_{n\to\infty} \dfrac{2n+1}{3n-2} \neq 1$.

(c) $\lim\limits_{n\to\infty} \dfrac{6}{\sqrt{n}} \neq .2$.

5.4 Questions for Reflection

In Section 5.3, you showed that if $\lim_{n\to\infty}(1/n)$ exists, then $\lim_{n\to\infty}(1/n) \neq .3$. You have not shown that the sequence has no limit, merely that the limit cannot be .3. In fact, this particular sequence does have a limit, namely 0. Yet not all sequences have a finite limit. The questions posed in this section are designed to help you to think about how to use the negation of the definition of convergence to prove that a given sequence has NO finite limit.

Sequence	$\epsilon = .5$	$\epsilon = .4$	$\epsilon = .3$	$\epsilon = .1$
$(a_n)_{n=1}^{\infty}$				
$(b_n)_{n=1}^{\infty}$				
$(c_n)_{n=1}^{\infty}$				

1. For a given sequence, how does the problem of showing that the sequence has no finite limit differ from showing that a particular real number is not the limit?

2. Define $(a_n)_{n=1}^{\infty} = \left(\begin{cases} 0, & \text{if } n \text{ is even} \\ 1, & \text{if } n \text{ is odd} \end{cases} \right)_{n=1}^{\infty}$.

   (a) Prove that if $\lim_{n \to \infty} a_n$ exists, then $\lim_{n \to \infty} a_n \neq 1$.

   (b) As you may have observed, this sequence has no limit. If you were to prove this claim following the plan suggested by your answer to question 1, you might find the procedure infinitely long or very cumbersome. Thus, another approach might be easier. Can you suggest an alternate approach?

   (c) Let L be any real number. Apply the method suggested by your previous answer to prove that if $\lim_{n \to \infty} a_n$ exists, then $\lim_{n \to \infty} a_n \neq L$. In this way, you have shown that the sequence diverges.

3. Consider the sequence $(n^2)_{n=1}^{\infty}$.

   (a) Just as you did in the previous question, prove that the sequence $(n^2)_{n=1}^{\infty}$ has no finite limit.

   (b) There is an easier way to prove this sequence is not convergent. Recall that in Lab 4 you proved that every convergent sequence is bounded. Use this fact to prove that the given sequence is not convergent.

4. Is it true that every bounded sequence is convergent? If so, prove the statement. Otherwise give a counterexample.

5. Suppose that $(a_n)_{n=1}^{\infty}$ is a sequence of real numbers which converges to a number L. Let $M \in \mathbb{R}$ such that $M \neq L$.

   (a) Prove that there are infinitely many terms of the sequence (a_n) which are not equal to M.

   (b) In fact, one can prove a stronger statement: there exists a number N such that $n > N$ implies $a_n \neq M$. Prove the stronger statement.

6. In order to prove the stronger statement given in part (b) of question 5, it was important to know that the limit of the sequence $(a_n)_{n=1}^{\infty}$ exists. We show why that hypothesis is needed in the following example. Let

$$a_n = \begin{cases} \dfrac{n+1}{n+8}, & \text{if } n \text{ is even} \\ 0, & \text{if } n \text{ is odd} \end{cases}.$$

   (a) Prove that the sequence $(a_n)_{n=1}^{\infty}$ diverges.

   (b) Prove that $a_n \neq 0$ for infinitely many values of n.

   (c) Prove that there exists no number N such that $n > N$ implies $a_n \neq 0$.

6

Algebraic Combinations of Sequences

6.1 Introduction

As you know, analysis is the study of functions. The goal of analysis is to understand the behavior of functions using the special tools of the limit, the derivative, and the integral. Functions can be combined in a variety of ways. They can be added, subtracted, multiplied, or divided. Mathematicians are interested in knowing if a property or a behavior shared by two functions is preserved when the functions are combined algebraically. You considered these types of questions in calculus. For example, you were asked: Is the sum of differentiable functions differentiable? Is the product of differentiable functions differentiable?

In this lab, we consider similar questions involving sequences. We are interested in understanding the relationship between the behavior of an algebraic combination of two sequences and the individual sequences that make up the combination. In particular, what condition guarantees the convergence of a sum of two sequences? Of a product? Of a quotient? You will study a variety of examples in order to formulate conjectures involving sums, products, and quotients. After having considered a proof of a special case, you will provide a proof of each conjecture. This will require you to use the definition of the limit of a sequence you derived in Lab 3.

The proof techniques that are demonstrated in this lab will help you throughout your study of real analysis. Writing proofs using the $\epsilon - N$ definition of convergence is required in many situations. Hence, another purpose of this lab is to guide you in developing your skill in writing proofs that require the use of technical definitions involving limits.

For this lab, you may wish to use the accompanying *Visual Guide for Lab 6*, which can be found at www.saintmarys.edu/~jsnow. The *Visual Guide* for this section also appears in the appendix. Alternatively, you may use the *Maplet for Lab 6,* which can be found at www.saintmarys.edu/~jsnow.

6.2 The Sum of Two Convergent Sequences

6.2.1 Formulating a Conjecture

For each pair of sequences $(a_n)_{n=1}^{\infty}$ and $(b_n)_{n=1}^{\infty}$ given in the table, determine whether $(a_n+b_n)_{n=1}^{\infty}$ converges. In columns 2 and 3, record the limit of convergence of the individual sequences; if

Set	$\lim\limits_{n\to\infty} a_n$	$\lim\limits_{n\to\infty} b_n$	$(a_n + b_n)$	$\lim\limits_{n\to\infty} (a_n + b_n)$
Set I $$(a_n)_{n=1}^{\infty} = \left(\frac{1}{n}\right)_{n=1}^{\infty}$$ $$(b_n)_{n=1}^{\infty} = \left(\frac{5n-2}{n+4}\right)_{n=1}^{\infty}$$				
Set II $$(a_n)_{n=1}^{\infty} = \left(\frac{1-2n}{n+1}\right)_{n=1}^{\infty}$$ $$(b_n)_{n=1}^{\infty} = \left(2 - \frac{1}{n^2}\right)_{n=1}^{\infty}$$				
Set III $$(a_n)_{n=2}^{\infty} = \left(\sin\left(\frac{n\pi}{2}\right)\right)_{n=2}^{\infty}$$ $$(b_n)_{n=2}^{\infty} = \left(\frac{1}{\ln n}\right)_{n=2}^{\infty}$$				
Set IV $$(a_n)_{n=1}^{\infty} = \left(1 + \frac{2}{n}\right)_{n=1}^{\infty}$$ $$(b_n)_{n=1}^{\infty} = \left(\begin{cases} 3 - 1/n, & \text{if } n \text{ is even} \\ 3, & \text{if } n \text{ is odd} \end{cases}\right)_{n=1}^{\infty}$$				
Set V $$(a_n)_{n=1}^{\infty} = ((-1)^n)_{n=1}^{\infty}$$ $$(b_n)_{n=1}^{\infty} = ((-1)^{n+1})_{n=1}^{\infty}$$				
Set VI $$(a_n)_{n=1}^{\infty} = (1^n)_{n=1}^{\infty}$$ $$(b_n)_{n=1}^{\infty} = ((-1)^n)_{n=1}^{\infty}$$				

the sequence does not converge, write DNE. In column 4, provide the general term of the sum. In column 5, give the limit of convergence; if the sum does not converge, write DNE.

Conjecture 1: Based upon the information given in the table, try to formulate a conjecture. In other words, what condition is sufficient to guarantee the convergence of the sum of two sequences? Write your conjecture in the form of an if-then statement.

6.2.2 Constructing a Proof

Let us outline a proof of Conjecture 1 by working with Set I:

$$(a_n)_{n=1}^{\infty} = \left(\frac{1}{n}\right)_{n=1}^{\infty} \qquad (b_n)_{n=1}^{\infty} = \left(\frac{5n-2}{n+4}\right)_{n=1}^{\infty}.$$

Discussion: Before continuing, let's review the definition of the convergence of a sequence. A sequence $(s_n)_{n=1}^{\infty}$ converges to real number L, that is, $\lim_{n\to\infty} s_n = L$, if and only if given $\epsilon > 0$ there exists a number N such that if $n > N$, then $|s_n - L| < \epsilon$.

As you learned in Lab 3, the definition incorporates our intuitive notion of convergence — as n grows large, the terms of the sequence approach the limiting value L — by considering the dependence of N upon ϵ. As a freely selected value, ϵ indicates the degree of closeness, while the ability to find a suitable N determines whether all terms beyond N satisfy that indicated degree of closeness. In this sense, the definition reads: in order for a sequence to converge, a suitable N must be found for each and every indicated degree of closeness. With regard to the sum of convergent sequences, we use the sequences in Step I to outline the proof showing that the sum of two convergent sequences is itself convergent; its limit is the sum of the limits of convergence of each of its summands.

In order to simplify things, we consider the process of finding N for a specified value of ϵ. You will be asked to consider the more general case later.

Let $\epsilon = .2$. We want to find a number N such that

$$\text{if} \quad n > N, \quad \text{then} \quad \left|\left(\frac{1}{n} + \frac{5n-2}{n+4}\right) - (0+5)\right| < \epsilon.$$

A natural way to do this is to isolate the pieces about which we know something. Since each summand is convergent, we take the expression

$$\left|\left(\frac{1}{n} + \frac{5n-2}{n+4}\right) - (0+5)\right|,$$

and separate each summand. This will eventually allow us to apply the definition to each component:

$$\left|\left(\frac{1}{n} + \frac{5n-2}{n+4}\right) - (0+5)\right| = \left|\left(\frac{1}{n} - 0\right) + \left(\frac{5n-2}{n+4} - 5\right)\right|$$

$$\leq \left|\frac{1}{n} - 0\right| + \left|\frac{5n-2}{n+4} - 5\right|.$$

Since we want the expression

$$\left|\left(\frac{1}{n} + \frac{5n-2}{n+4}\right) - (0+5)\right|$$

to be less than ϵ, we would like each of the summands,

$$\left|\frac{1}{n} - 0\right| \quad \text{and} \quad \left|\frac{5n-2}{n+4} - 5\right|,$$

to be less than $\epsilon/2$. Since the sequence $\left(\frac{1}{n}\right)_{n=1}^{\infty}$ converges, we can find a number N_1 beyond which all terms will be within $\epsilon/2$ units of 0; that is,

$$\text{if} \quad n > N_1, \quad \text{then} \quad \left|\frac{1}{n} - 0\right| < .1. \tag{5}$$

Similarly, since the sequence $\left(\frac{5n-2}{n+4}\right)_{n=1}^{\infty}$ converges, we can find a number N_2 beyond which all terms are within $\epsilon/2$ units of $5/2$; that is,

$$\text{if} \quad n > N_2, \quad \text{then} \quad \left|\frac{5n-2}{n+4} - 5\right| < .1. \tag{6}$$

What values will work for N_1? for N_2? You answered these questions in Lab 4. Go back to this lab, and find these values for N_1 and N_2. What value of N should we use to make both (1) and (2) true? Why is it important to make sure that both (1) and (2) hold?

Based upon your answers to the last two questions, complete the proof given below for the case in which $\epsilon = .2$, and then justify each step of the proof.

Proof. Let $\epsilon = .2$. Let $N =$ (determine a formula for N). Then for $n > N$,

$$\left|\left(\frac{1}{n} + \frac{5n-2}{n+4}\right) - (0 + 5)\right| = \left|\left(\frac{1}{n} - 0\right) + \left(\frac{5n-2}{n+4} - 5\right)\right|$$

$$\leq \left|\frac{1}{n} - 0\right| + \left|\frac{5n-2}{n+4} - 5\right|$$

$$< .1 + .1 = .2.$$

Q.E.D.

6.2.3 The Conjecture and Its Proof

1. Using the proof in Section 6.2.2 as a guide, provide a formal proof of Conjecture 1.

2. The "if" part of Conjecture 1 gives a condition which guarantees the convergence of the sum of two sequences. Based upon the examples you considered, is it possible for the sum of two sequences to converge without this condition being satisfied? If so, does this invalidate the statement of your conjecture? What implication, if any, does this have for the converse of Conjecture 1?

6.3 The Product of Two Convergent Sequences

6.3.1 Formulating a Conjecture

For each pair of sequences $(a_n)_{n=1}^{\infty}$ and $(b_n)_{n=1}^{\infty}$ given in the table, determine whether $(a_n \cdot b_n)_{n=1}^{\infty}$ converges. In columns 2 and 3, record the limit of convergence of the individual sequences; if the limit of convergence does not exist, write DNE. In column 4, provide the general term of the product. In column 5, give the limit of convergence of the product; if the product does not converge, write DNE.

Set	$\lim\limits_{n \to \infty} a_n$	$\lim\limits_{n \to \infty} b_n$	$(a_n \cdot b_n)$	$\lim\limits_{n \to \infty} (a_n \cdot b_n)$
Set I $$(a_n)_{n=1}^{\infty} = \left(\frac{2+3n}{n+4}\right)_{n=1}^{\infty}$$ $$(b_n)_{n=1}^{\infty} = \left(\frac{5n-2}{n+4}\right)_{n=1}^{\infty}$$				
Set II $$(a_n)_{n=1}^{\infty} = (n)_{n=1}^{\infty}$$ $$(b_n)_{n=1}^{\infty} = \left(\frac{1}{n^2}\right)_{n=1}^{\infty}$$				
Set III $$(a_n)_{n=1}^{\infty} = \left(3^{\frac{1}{n}}\right)_{n=1}^{\infty}$$ $$(b_n)_{n=1}^{\infty} = \left(2^{\frac{1}{n^2}}\right)_{n=1}^{\infty}$$				
Set IV $$(a_n)_{n=1}^{\infty} = \left(3 + \frac{1}{n}\right)_{n=1}^{\infty}$$ $$(b_n)_{n=1}^{\infty} = \left(\begin{cases} 2 - \dfrac{1}{n^2}, & \text{if } n \text{ is even} \\ 2, & \text{if } n \text{ is odd} \end{cases}\right)_{n=1}^{\infty}$$				
Set V $$(a_n)_{n=1}^{\infty} = \left(\sin\left(\frac{n\pi}{2}\right)\right)_{n=1}^{\infty}$$ $$(b_n)_{n=1}^{\infty} = ((-1)^n)_{n=1}^{\infty}$$				

Conjecture 2: Based upon the information given in the table, try to formulate a conjecture. In other words, what condition is sufficient to guarantee the convergence of the product of two sequences? Write the conjecture in the form of an if-then statement.

6.3.2 Constructing a Proof

Let us outline a proof of Conjecture 2 by working with Set III:

$$(a_n)_{n=1}^{\infty} = \left(3^{1/n}\right)_{n=1}^{\infty} \qquad (b_n)_{n=1}^{\infty} = \left(2^{1/n^2}\right)_{n=1}^{\infty}.$$

Discussion: Similar to the case involving the sum, we will outline the proof for a specific value of ϵ (in this case, $\epsilon = .2$), and then proceed to find a value of N such that

$$\text{if} \quad n > N, \quad \text{then} \quad \left|\left(3^{1/n} \cdot 2^{1/n^2}\right) - (1 \cdot 1)\right| < \epsilon.$$

Like the discussion in Section 6.2.2, we wish to isolate those terms about which we know something. In particular, we use the fact that each individual sequence converges to prove convergence of the product. However, unlike the sum, there is no obvious way to split the expression $\left|\left(3^{1/n} \cdot 2^{1/n^2}\right) - (1 \cdot 1)\right|$ into an algebraic combination of $\left|3^{1/n} - 1\right|$ and $\left|2^{1/n^2} - 1\right|$.

One way to deal with this problem is to modify the original expression. In this case, this means adding and then subtracting $3^{1/n}$. This will allow us to apply the distributive property to write a sum in which the terms $\left|3^{1/n} - 1\right|$ and $\left|2^{1/n^2} - 1\right|$ appear:

$$\left|\left(3^{1/n} \cdot 2^{1/n^2}\right) - 3^{1/n} + 3^{1/n} - 1\right| = \left|\left[3^{1/n}\left(2^{1/n^2} - 1\right)\right] + \left[3^{1/n} - 1\right]\right|$$

$$\leq \left|3^{1/n}\right|\left|2^{1/n^2} - 1\right| + \left|3^{1/n} - 1\right|. \tag{$*$}$$

The problem is the presence of the nonconstant term, $\left|3^{1/n}\right|$. However, since $\left(3^{1/n}\right)_{n=1}^{\infty}$ is a positive-term, decreasing sequence that converges to 1, it is quite easy to see that

$$\left|3^{1/n}\right| < 4.$$

For the general proof, which you consider in Section 6.3.3, the issue of constructing a bound for the "extra" term involves direct application of the definition of convergence: a specified $\epsilon > 0$ is chosen, and the definition is applied to find a suitable N. The reason for selecting a specific value of ϵ is that it simplifies the proof.

Since we want the entire sum to be less than ϵ, we take each summand to be less than $\epsilon/2$. Since the sequence $\left(3^{1/n}\right)_{n=1}^{\infty}$ converges to 1, we can find a number N_1 such that

$$\text{if} \quad n > N_1, \quad \text{then} \quad \left|3^{1/n} - 1\right| < \frac{.2}{2}. \tag{1}$$

At this point, find a value for N_1.

For the second factor of the first summand in $(*)$, we can use the definition to find a number N_2 such that

$$\text{if} \quad n > N_2, \quad \text{then} \quad \left|2^{1/n^2} - 1\right| < \frac{.2}{8}. \tag{2}$$

Find a suitable N_2.

Can you name a number N that guarantees that (1) and (2) hold simultaneously? Why is it important to make sure that we can find such an N?

Based upon your answers to the last two questions, complete the proof given below for the case in which $\epsilon = .2$, and then justify each step of the proof.

Proof. Let $\epsilon = .2$. Let $N =$ (determine a formula for N). Then for $n > N$,

$$\left|\left(3^{1/n} \cdot 2^{1/n^2}\right) - 3^{1/n} + 3^{1/n} - 1\right| = \left|\left[3^{1/n}\left(2^{1/n^2} - 1\right)\right] + \left(3^{1/n} - 1\right)\right|$$

$$\leq \left|3^{1/n}\right|\left|2^{1/n^2} - 1\right| + \left|3^{1/n} - 1\right|$$

$$< (4)\frac{.2}{8} + \frac{.2}{2} = .2 \qquad \qquad \text{Q.E.D.}$$

6.3.3   The Conjecture and Its Proof

1. Using the proof in Section 6.3.2 as a guide, provide a formal proof of Conjecture 2.

2. The "if" part of Conjecture 2 gives a condition that guarantees the convergence of the product of two sequences. Based upon the examples you considered, is it possible for the product of two sequences to converge without this condition being satisfied? If so, does this invalidate the statement of your conjecture? What implication, if any, does this have for the converse of Conjecture 2?

6.4   The Quotient of Two Convergent Sequences

6.4.1   Formulating a Conjecture

For each sequence $(a_n)_{n=1}^{\infty}$, determine whether $(1/a_n)_{n=1}^{\infty}$ converges. In column 2, record the limit of the individual sequence; if the sequence does not converge, write DNE. In column 3, provide the general term of the reciprocal. In column 4, give the limit of the reciprocal; if the reciprocal sequence does not converge, write DNE.

Sequence	$\lim_{n\to\infty} a_n$	$(1/a_n)$	$\lim_{n\to\infty} (1/a_n)$
$\left(3 - \dfrac{2}{n}\right)_{n=1}^{\infty}$			
$\left(\sin\left(\dfrac{n\pi}{2}\right)\right)_{n=1}^{\infty}$			
$\left(\dfrac{2n}{3n+4}\right)_{n=1}^{\infty}$			
$\left(\dfrac{1}{n^2}\right)_{n=1}^{\infty}$			
$\left(\begin{cases} 3 - \dfrac{1}{n}, & \text{if } n \text{ is even} \\ 3, & \text{if } n \text{ is odd} \end{cases}\right)_{n=1}^{\infty}$			
$\left(\dfrac{n!}{25^n}\right)_{n=1}^{\infty}$			

Conjecture 3. Based upon the information given in the table, try to formulate a conjecture. In other words, what condition is sufficient to guarantee the convergence of the reciprocal of a sequence? Write the conjecture in the form of an if-then statement.

6.4.2   Constructing a Proof

In this case, we proceed a bit differently than we did in Sections 6.2.2 and 6.3.2. Rather than work with the quotient of two specific sequences, we consider the reciprocal of a single arbitrary sequence and deal with the issue of the quotient later. Specifically, if we assume $(a_n)_{n=1}^{\infty}$ converges to $L \neq 0$, we want to show that $(1/a_n)_{n=1}^{\infty}$ converges to $1/L$. Why is L required to be nonzero? Is this also a requirement for the terms a_n? Why, or why not?

Discussion.   In this situation, we discuss the general case. Let ϵ denote a positive real number. We want to find a number N such that

$$\text{if } n > N, \text{ then } \left| \frac{1}{a_n} - \frac{1}{L} \right| < \epsilon.$$

Similar to the situations involving the sum and product, we use the fact that the sequence $(a_n)_{n=1}^{\infty}$ converges. This means that we need to be able to modify or to rewrite the expression

$$\left| \frac{1}{a_n} - \frac{1}{L} \right| \tag{$*$}$$

so that we can apply the definition to the term $|a_n - L|$. As it turns out, we can do this quite easily by finding a common denominator for $(*)$:

$$\left| \frac{1}{a_n} - \frac{1}{L} \right| = \left| \frac{L - a_n}{a_n \cdot L} \right|$$

$$= \frac{|a_n - L|}{|L|\,|a_n|}.$$

Although we obtain an expression with the term $|a_n - L|$, we have introduced a nonconstant term that needs to be bounded.

In finding a suitable bound, we must be aware that we are actually working with the reciprocal of $|a_n|$. Similar to the strategy followed in Section 6.3.2, we fix a value of ϵ. In this case, let $\epsilon = \frac{|L|}{2}$. This will ensure that the bound on $\frac{1}{|a_n|}$ is not a variable expression in terms of ϵ. Since $(a_n)_{n=1}^{\infty}$ converges to L, there exists a number, say N_1, such that

$$\text{if } n > N_1, \text{ then } |a_n - L| < \frac{|L|}{2}.$$

We will use this inequality to place an upper bound on the reciprocal:

$$|L| = |L - a_n + a_n|$$

$$\leq |L - a_n| + |a_n|$$

$$= |a_n - L| + |a_n|$$

$$< \frac{|L|}{2} + |a_n|, \quad \text{provided } n > N_1.$$

Complete the steps to show that

$$\text{if } n > N_1, \text{ then } \frac{1}{|a_n|} < \frac{2}{|L|}. \tag{1}$$

Now that we have found a bound for $1/|a_n|$, we can consider the term $|a_n - L|$. Since the sequence $(a_n)_{n=1}^{\infty}$ converges, we can find a number, say N_2, such that

$$\text{if} \quad n > N_2, \quad \text{then} \quad |a_n - L| < \frac{|L|^2 \epsilon}{2}. \tag{2}$$

Why do we work with such a seemingly complicated expression involving ϵ in (2)?

   Hint: Remember that the goal is to find N such that if $n > N$ then

$$\frac{|a_n - L|}{|a_n||L|} < \epsilon.$$

   The statements (1) and (2) given above form the basis for a proof, which you are asked to write in the next section.

6.4.3 The Conjecture and Associated Proofs

1. Use the discussion given in Section 6.4.2 to help you write a formal proof of Conjecture 3.

2. Let $(a_n)_{n=1}^{\infty}$ be a sequence that converges to L. Let $(b_n)_{n=1}^{\infty}$ be a nonzero sequence that converges to M, where $M \neq 0$. Prove that the quotient

$$\left(\frac{a_n}{b_n}\right)_{n=1}^{\infty}$$

    converges to L/M.

3. The "if" part of Conjecture 3 gives a condition which guarantees the convergence of the reciprocal of a sequence. Based upon the examples you considered, is it possible for the reciprocal of a sequence to converge without this condition being satisfied? If so, does this invalidate the statement of your conjecture? What implication, if any, does this have for the converse of Conjecture 3?

6.5 Weakening the Condition Involving Products

Is it possible to weaken the condition in Conjecture 2? This differs from the questions posed in Sections 6.2.3, 6.3.3, and 6.4.3 regarding converses. In this case, you are considering the statement of Conjecture 2 itself; you are being asked whether the condition requiring the convergence of each sequence can be weakened. In other words, can we find a condition that guarantees the convergence of the product of two sequences without necessarily requiring that both sequences converge? To help answer this question, consider the following pairs of sequences provided in the table that follows. In columns 2 and 3, record the limit of convergence; if the sequence does not converge, write DNE. In column 4, provide the general term of the product of the two sequences. In column 5, give the limit of convergence; if the product does not converge, write DNE.

1. For each set given here, what is $\lim\limits_{n \to \infty} (a_n b_n)$?

2. What common properties do the sequences $(a_n)_{n=1}^{\infty}$ have?

3. What common properties do the sequences $(b_n)_{n=1}^{\infty}$ have?

Set	$\lim_{n\to\infty} a_n$	$\lim_{n\to\infty} b_n$	$c_n = a_n b_n$	$\lim_{n\to\infty} c_n$
Set I $(a_n)_{n=1}^{\infty} = \left(\dfrac{1}{n}\right)_{n=1}^{\infty}$ $(b_n)_{n=1}^{\infty} = ((-1)^n)_{n=1}^{\infty}$				
Set II $(a_n)_{n=1}^{\infty} = \left(\dfrac{1}{n^2}\right)_{n=1}^{\infty}$ $(b_n)_{n=1}^{\infty} = (\sin n)_{n=1}^{\infty}$				
Set III $(a_n)_{n=1}^{\infty} = \left(\dfrac{4}{n^3}\right)_{n=1}^{\infty}$ $(b_n)_{n=1}^{\infty} = \left(\begin{cases} 2 - \frac{1}{n}, & \text{if } n \text{ is even} \\ -2 + \frac{1}{n}, & \text{if } n \text{ is odd} \end{cases}\right)_{n=1}^{\infty}$				
Set IV $(a_n)_{n=1}^{\infty} = \left(\dfrac{(-1)^n}{n}\right)_{n=1}^{\infty}$ $(b_n)_{n=1}^{\infty} = \left(\begin{cases} 2, & \text{if } n \text{ is even} \\ \frac{1}{n}, & \text{if } n \text{ is odd} \end{cases}\right)_{n=1}^{\infty}$				

4. Formulate a conjecture (Conjecture 4) that reflects your findings.

5. Does Conjecture 4, if true, invalidate Conjecture 2?

6.6 Questions for Reflection

1. The sequence $(e^{-n})_{n=1}^{\infty}$ converges, but the sequence

$$\left(\frac{1}{e^{-n}}\right)_{n=1}^{\infty}$$

does not. What aspect of Conjecture 3 is being violated?

2. If $(a_n)_{n=1}^{\infty}$ is a convergent sequence, show that $(a_n)_{n=1}^{\infty}$ is bounded.

3. Provide a proof of Conjecture 4.

4. Let $\left(p_n\right)_{n=1}^{\infty} = \left(a_0 + a_1 n + \cdots + a_k n^k\right)_{n=1}^{\infty}$ and $\left(q_n\right)_{n=1}^{\infty} = \left(b_0 + b_1 n + \cdots + b_m n^m\right)_{n=1}^{\infty}$ be two sequences, where k and m are positive integers. Under what conditions does the quotient sequence

$$\left(\frac{p_n}{q_n}\right)_{n=1}^{\infty}$$

converge? Write a statement, and provide a proof.

5. Suppose that the sequence $\left(a_n\right)_{n=1}^{\infty}$ converges to a nonzero real number, and assume that the sequence $\left(b_n\right)_{n=1}^{\infty}$ is bounded. Determine whether the sequence $\left(c_n\right)_{n=1}^{\infty}$, where

$$c_n = \frac{a_n b_n + 5n}{a_n^3 + n},$$

converges. If so, provide a detailed proof.

6. Determine whether the following statements are true or false. For those statements that are false, provide a counterexample. For those that are true, provide a proof.

   (a) If $\left(a_n\right)_{n=1}^{\infty}$ is a sequence such that $\lim_{n \to \infty} a_n = \infty$ (as n increases, the terms a_n increase positively without bound), and if $\left(b_n\right)_{n=1}^{\infty}$ is a sequence that converges to 0, then $\left(a_n b_n\right)_{n=1}^{\infty}$ converges.

   (b) If $\left(a_n\right)_{n=1}^{\infty}$ is sequence that is positive and bounded, and if $\left(b_n\right)_{n=1}^{\infty}$ is a sequence of exclusively positive terms such that $\lim_{n \to \infty} b_n = \infty$, then

$$\left(\frac{a_n}{b_n}\right)_{n=1}^{\infty}$$

   converges.

   (c) If $\left(a_n\right)_{n=1}^{\infty}$ and $\left(b_n\right)_{n=1}^{\infty}$ are both positive-term sequences such that $\lim_{n \to \infty} a_n = \infty$ and $\lim_{n \to \infty} b_n = \infty$, then

$$\left(\frac{a_n}{b_n}\right)_{n=1}^{\infty}$$

   diverges.

7

![chapter heading bar]

Conditions Related to Convergence

7.1 Introduction

In this lab, we continue our study of convergence of a sequence of real numbers. Now we examine two issues: conditions which are consequences of convergence and conditions which yield convergence. If a sequence does converge, we can make a strong statement about the range and variability of its terms. In this lab you discover that statement. Another task is to find a pair of conditions that guarantee convergence. Finally, you are asked to determine a condition that is equivalent to convergence. It is useful to have alternate approaches to prove convergence, because our knowledge about a given sequence is not always the same.

You will need the following definitions.

- A sequence $(s_n)_{n=1}^{\infty}$ is said to be *monotonic* if either

$$s_n \leq s_{n+1} \quad \text{or} \quad s_n \geq s_{n+1} \text{ for all } n \in \mathbb{N}.$$

- A sequence $(s_n)_{n=1}^{\infty}$ is said to be *Cauchy* if for every $\epsilon > 0$, there exists a number N such that $n, m > N$ implies $|s_n - s_m| < \epsilon$.

Loosely speaking, a sequence is Cauchy if the difference between the terms of the sequence tends toward 0. The terms which are compared need not be sequential.

7.2 Using Examples to Enhance Understanding

As a first step towards the goals outlined above, you are asked to gather some information for some example sequences. For each sequence, answer the following set of questions and record your answers in the appropriate column of the accompanying table. The questions are:

1. Is the sequence convergent? Record the answer (yes or no) in column Q1.

2. Is the sequence bounded? Record the answer (yes or no) in column Q2.

3. Is the sequence monotonic? Record the answer (yes or no) in column Q3.

4. Is the sequence Cauchy? To answer this question, compute the absolute value of the difference of some terms. That is, compute $|a_{50} - a_{25}|$, $|a_{100} - a_{80}|$, $|a_{225} - a_{175}|$, $|a_{500} - a_{400}|$, and $|a_{850} - a_{825}|$. Are the differences tending toward zero as both subscripts tend toward infinity? Record your answer to this last question (just a yes or no) in column Q4.

Sequence 1: $(1/n)_{n=1}^{\infty}$

Q1	Q2	Q3	$|a_{50} - a_{25}|$	$|a_{100} - a_{80}|$	$|a_{225} - a_{175}|$	$|a_{500} - a_{400}|$	$|a_{850} - a_{825}|$	Q4

Sequence 2: $\left(\begin{cases} 2 + \dfrac{1}{n}, & \text{if } n \text{ is even} \\ -2 - \dfrac{1}{n}, & \text{if } n \text{ is odd} \end{cases} \right)_{n=1}^{\infty}$

Q1	Q2	Q3	$|a_{50} - a_{25}|$	$|a_{100} - a_{80}|$	$|a_{225} - a_{175}|$	$|a_{500} - a_{400}|$	$|a_{850} - a_{825}|$	Q4

Sequence 3: $\left(\dfrac{5n - 2}{n + 4} \right)_{n=1}^{\infty}$

Q1	Q2	Q3	$|a_{50} - a_{25}|$	$|a_{100} - a_{80}|$	$|a_{225} - a_{175}|$	$|a_{500} - a_{400}|$	$|a_{850} - a_{825}|$	Q4

Sequence 4: $(\sin(n))_{n=1}^{\infty}$

Q1	Q2	Q3	$|a_{50} - a_{25}|$	$|a_{100} - a_{80}|$	$|a_{225} - a_{175}|$	$|a_{500} - a_{400}|$	$|a_{850} - a_{825}|$	Q4

Sequence 5: $\left(\begin{cases} 3 - e^{-n}, & \text{if } n \text{ is even} \\ 3, & \text{if } n \text{ is odd} \end{cases} \right)_{n=1}^{\infty}$

Q1	Q2	Q3	$|a_{50} - a_{25}|$	$|a_{100} - a_{80}|$	$|a_{225} - a_{175}|$	$|a_{500} - a_{400}|$	$|a_{850} - a_{825}|$	Q4

Sequence 6: $\left(2 + \dfrac{(-1)^n}{n} \right)_{n=1}^{\infty}$

Q1	Q2	Q3	$|a_{50} - a_{25}|$	$|a_{100} - a_{80}|$	$|a_{225} - a_{175}|$	$|a_{500} - a_{400}|$	$|a_{850} - a_{825}|$	Q4

Sequence 7: $(\cos(n\pi/2))_{n=1}^{\infty}$

Q1	Q2	Q3	$|a_{50} - a_{25}|$	$|a_{100} - a_{80}|$	$|a_{225} - a_{175}|$	$|a_{500} - a_{400}|$	$|a_{850} - a_{825}|$	Q4

Sequence 8: $(n)_{n=1}^{\infty}$

Q1	Q2	Q3	$|a_{50} - a_{25}|$	$|a_{100} - a_{80}|$	$|a_{225} - a_{175}|$	$|a_{500} - a_{400}|$	$|a_{850} - a_{825}|$	Q4

Sequence 9: $\left(\sqrt{4 + \dfrac{20}{n}} \right)_{n=1}^{\infty}$

Q1	Q2	Q3	$|a_{50} - a_{25}|$	$|a_{100} - a_{80}|$	$|a_{225} - a_{175}|$	$|a_{500} - a_{400}|$	$|a_{850} - a_{825}|$	Q4

Sequence 10: $\left(\begin{cases} 4^n, & \text{if } n \text{ is even} \\ 4^{-n}, & \text{if } n \text{ is odd} \end{cases} \right)_{n=1}^{\infty}$

Q1	Q2	Q3	$\lvert a_{50} - a_{25} \rvert$	$\lvert a_{100} - a_{80} \rvert$	$\lvert a_{225} - a_{175} \rvert$	$\lvert a_{500} - a_{400} \rvert$	$\lvert a_{850} - a_{825} \rvert$	Q4

Sequence 11: $\left(\dfrac{(-1)^n}{n^2} \right)_{n=1}^{\infty}$

Q1	Q2	Q3	$\lvert a_{50} - a_{25} \rvert$	$\lvert a_{100} - a_{80} \rvert$	$\lvert a_{225} - a_{175} \rvert$	$\lvert a_{500} - a_{400} \rvert$	$\lvert a_{850} - a_{825} \rvert$	Q4

Sequence 12: $\left(e^n \right)_{n=1}^{\infty}$

Q1	Q2	Q3	$\lvert a_{50} - a_{25} \rvert$	$\lvert a_{100} - a_{80} \rvert$	$\lvert a_{225} - a_{175} \rvert$	$\lvert a_{500} - a_{400} \rvert$	$\lvert a_{850} - a_{825} \rvert$	Q4

7.3   Critical Thinking Questions

Use the information you have gathered in Section 7.2 to answer the following questions.
First, we will focus on consequences of convergence.

1. Does every convergent sequence seem to be bounded? If not, give a counterexample.

2. Does every convergent sequence seem to be monotonic? If not, give a counterexample.

3. Does every convergent sequence seem to be Cauchy? If not, give a counterexample.

4. Based on your answers to questions 1–3, which of the following properties seem to be consequences of convergence: boundedness, monotonicity, and/or Cauchy? That is, can any of the properties be used to complete the following: "If a sequence is convergent, then it is . . ."?

Now let us look for conditions which yield convergence.

5. Does it appear that every bounded sequence is convergent? If not, give a counterexample.

6. Does it appear that every monotonic sequence is convergent? If not, give a counterexample.

7. Does it appear that every Cauchy sequence is convergent? If not, give a counterexample.

8. Based on your answers to questions 5–7, do you see any condition which is sufficient for convergence; that is, if a sequence has this property, then it converges?

9. Now phrase your answer to question 8 in terms of an "if ..., then ..." statement. That is, complete the following sentence: "If a sequence is ..., then the sequence converges."

10. You may have observed that some of the conditions in questions 5–7 did not yield convergence. Sometimes, however, if a sequence possesses a combination of properties, convergence results. Do you see any combination of conditions which yield convergence? That is, complete the following sentence: "If a sequence is ... and ..., then the sequence converges."

11. Reviewing your answers to questions 4 and 9, do you find any condition which is equivalent to convergence? That is, can you complete the following sentence: "A sequence is convergent if and only if ...."

7.4 Questions for Reflection

1. You have generated a condition that is equivalent to convergence. That means that either the definition or the equivalent condition can be used to prove convergence. In what situations does the definition serve as a useful tool? Under what circumstances would it be useful to use the equivalent condition you have formulated in this lab?

2. Let $.a_1 a_2 a_3 \ldots$ be an infinite decimal, where a_i is an integer between 0 and 9. Approximations to this decimal can form a sequence:

$$.a_1, \quad .a_1 a_2, \quad .a_1 a_2 a_3, \ldots.$$

Using the results you found in Section 3, prove that the sequence of approximations converges.

3. Let $(a_n)_{n=1}^{\infty}$ be a sequence such that $a_1 = 1$ and

$$a_n = \frac{2n+1}{3n} a_{n-1}$$

for $n > 1$. Prove that $(a_n)_{n=1}^{\infty}$ converges.

4. Prove that every Cauchy sequence is bounded.

5. Prove that if an increasing, bounded sequence converges, then a nonempty subset of real numbers has a supremum. (Yes, we are asking you to prove that the completeness axiom holds under the given condition.) (Hint: Let S be the bounded set. There exist real numbers a and b such that $S \subset [a, b]$. Cut the interval $[a, b]$ into ten equal subintervals:

$$[\tilde{a}_1 = a, \tilde{b}_1], [\tilde{a}_2, \tilde{b}_2], [\tilde{a}_3, \tilde{b}_3], \ldots, [\tilde{a}_{10}, \tilde{b}_{10} = b].$$

Choose the subinterval which is the last one which contains points of the set; that is, choose the subinterval $[\tilde{a}_k, \tilde{b}_k]$ such that there are points of S in that subinterval, but no points of S in the subinterval $[\tilde{a}_{k+1}, \tilde{b}_{k+1}]$; that is, $\tilde{b}_k$ is an upper bound of S. Rename the interval $[\tilde{a}_k, \tilde{b}_k]$ as $[a_1, b_1]$. Repeat the process. Cut the interval $[a_1, b_1]$ into ten equal subintervals. Choose the subinterval which is the last one which contains points of the set; denote that interval as $[a_2, b_2]$. Again the right endpoint of this subinterval is an upper bound of the set. Repeat this process, thus producing a sequence $(a_n)_{n=1}^{\infty}$. Show this sequence converges and relate the limit to the supremum of the set. (Historical note: This is a process due to Karl Weierstrass.)

8

Understanding the Limit Superior and the Limit Inferior

8.1 Introduction

Given an *arbitrary* sequence $(a_n)_{n=1}^\infty$, we can construct two sequences from the original sequence such that each of these newly constructed sequences has a limit. (Notice the power of what has just been said. The original sequence need not have a limit, but each of the two sequences derived from it will have a limit!) The limits of the sequences we construct are called the *limit superior* (often denoted lim sup) and the *limit inferior* (often denoted lim inf). We can use these constructions to answer questions about the convergence of the original sequence. Later you will see that these constructions also help us determine the question of convergence of series.

Note that we allow $+\infty$ and $-\infty$ as the limit of a sequence. We use the stronger word *converge* when the limit of a sequence is a real number.

8.2 The Construction for Limit Superior

1. We begin construction of the first of the sequences derived from a given sequence $(a_n)_{n=1}^\infty$. In column 1 in the table that follows is a list of sequences. For each of these sequences, answer parts (a)–(d) and record your answers in the appropriate column in Table 8.1. Graphing $(a_n)_{n=1}^\infty$ may be helpful.

   (a) Find $v_1 := \sup\{a_n : n > 1\}$. Record your answer in column 2.

   (b) Find $v_{10} := \sup\{a_n : n > 10\}$. Record your answer in column 3.

   (c) Find $v_{100} := \sup\{a_n : n > 100\}$. Record your answer in column 4.

   (d) Find $v_{1000} := \sup\{a_n : n > 1000\}$. Record your answer in column 5.

2. As you probably have observed, the subscript values 1, 10, 100, and 1000 were arbitrarily chosen. For any natural number k, we define $v_k := \sup\{a_n : n > k\}$. If v_k is a real number

47

TABLE 8.1

Sequence	v_1	v_{10}	v_{100}	v_{1000}
$(1/n)_{n=1}^{\infty}$				
$\left(\cos(n\pi/2)\right)_{n=1}^{\infty}$				
$\left(2 + ((-1)^n/n)\right)_{n=1}^{\infty}$				
$((n+1)/n)_{n=1}^{\infty}$				
$\left(\begin{cases} 3 - e^{-n}, & \text{if } n \text{ is even} \\ 3, & \text{if } n \text{ is odd} \end{cases}\right)_{n=1}^{\infty}$				

for all k, we denote the sequence so defined by $(v_k)_{k=1}^{\infty}$. For each of the sequences in column 1 of Table 8.1, use the data you entered in answering question 1 to help you find a formula for v_k. Record your answer in column 2 of Table 8.2. (You will complete the last column of the table in question 6.)

3. Looking at the answers that you recorded in the tables, do you notice a pattern in the terms of $(v_k)_{k=1}^{\infty}$? That is, are the sequences $(v_k)_{k=1}^{\infty}$ bounded, monotonic, constant, non-increasing, or non-decreasing? Record any characteristics common to all the sequences.

4. Explain why the characteristics you recorded in question 3 would hold for $(v_k)_{k=1}^{\infty}$ regardless of the sequence $(a_n)_{n=1}^{\infty}$.

5. Why does $\lim_{k\to\infty} v_k$ exist (regardless of the sequence $(a_n)_{n=1}^{\infty}$)? (Recall that we allow ∞ or $-\infty$ as a valid limit.)

6. We write $\limsup a_n = \lim_{k\to\infty} v_k$. Use the formula you recorded in column 2 of Table 8.2 to help you determine $\limsup a_n$ for each sequence. Record your answer in column 3.

TABLE 8.2

Sequence	Formula for v_k	$\limsup(a_n) = \lim_{k\to\infty} v_k$
$(1/n)_{n=1}^{\infty}$		
$\left(\cos(n\pi/2)\right)_{n=1}^{\infty}$		
$\left(2 + ((-1)^n/n)\right)_{n=1}^{\infty}$		
$((n+1)/n)_{n=1}^{\infty}$		
$\left(\begin{cases} 3 - e^{-n}, & \text{if } n \text{ is even} \\ 3, & \text{if } n \text{ is odd} \end{cases}\right)_{n=1}^{\infty}$		

7. For the sequence

$$\left(a_n = \frac{2n + 11}{n + 2}\right)_{n=1}^{\infty},$$

find the first 6 terms of the sequence $(v_k)_{k=1}^{\infty}$ and find $\limsup a_n$.

8. For the sequence

$$\left(a_n = \frac{n - 1}{6n}\right)_{n=1}^{\infty},$$

find the first 6 terms of the sequence $(v_k)_{k=1}^{\infty}$ and find $\limsup a_n$.

9. For the sequence

$$\left(a_n = \begin{cases} 2 + \dfrac{1}{n}, & \text{if } n \text{ is even} \\ 0, & \text{if } n \text{ is odd} \end{cases}\right)_{n=1}^{\infty},$$

find the first 6 terms of the sequence $(v_k)_{k=1}^{\infty}$ and find $\limsup a_n$.

8.3 The Construction for Limit Inferior

1. We now construct the second sequence derived from a given sequence $(a_n)_{n=1}^{\infty}$. For each sequence in the table, record your answers to parts (a)–(d) in the appropriate column of Table 8.3. Again, graphing the sequences may be helpful.

(a) Find $u_1 := \inf\{a_n : n > 1\}$. Record your answer in column 2.

(b) Find $u_{10} := \inf\{a_n : n > 10\}$. Record your answer in column 3.

(c) Find $u_{100} := \inf\{a_n : n > 100\}$. Record your answer in column 4.

(d) Find $u_{1000} := \inf\{a_n : n > 1000\}$. Record your answer in column 5.

TABLE 8.3

Sequence	u_1	u_{10}	u_{100}	u_{1000}
$(1/n)_{n=1}^{\infty}$				
$(\cos(n\pi/2))_{n=1}^{\infty}$				
$(2 + ((-1)^n/n))_{n=1}^{\infty}$				
$((n+1)/n)_{n=1}^{\infty}$				
$\left(\begin{cases} 3 - e^{-n}, & \text{if } n \text{ is even} \\ 3, & \text{if } n \text{ is odd} \end{cases}\right)_{n=1}^{\infty}$				

2. Again as you probably have observed, the subscript values 1, 10, 100, and 1000 were arbitrarily chosen. For any natural number k, we can define $u_k := \inf\{a_n : n > k\}$. If u_k is a real number for all k, we denote the sequence so defined by $(u_k)_{k=1}^{\infty}$. For each of the sequences in column 1 of Table 8.3, use the data you entered in the table for question 1 to help you find a formula for u_k. Record your answer in column 2 of Table 8.4. (You will complete the last column of the table in question 6.)

TABLE 8.4

Sequence	Formula for u_k	$\liminf(a_n) = \lim\limits_{k\to\infty} u_k$
$(1/n)_{n=1}^{\infty}$		
$(\cos(n\pi/2))_{n=1}^{\infty}$		
$(2 + ((-1)^n/n))_{n=1}^{\infty}$		
$((n+1)/n)_{n=1}^{\infty}$		
$\left(\begin{cases} 3 - e^{-n}, & \text{if } n \text{ is even} \\ 3, & \text{if } n \text{ is odd} \end{cases}\right)_{n=1}^{\infty}$		

3. Looking at the answers that you recorded in the two preceding tables, do you notice a pattern in the terms of $(u_k)_{k=1}^{\infty}$? That is, are the sequences $(u_k)_{k=1}^{\infty}$ bounded, monotonic, constant, non-increasing, or non-decreasing? Record any characteristics common to all the sequences.

4. Explain why the characteristics you recorded in question 3 would hold for $(u_k)_{k=1}^{\infty}$ regardless of the sequence $(a_n)_{n=1}^{\infty}$.

5. Why does $\lim_{k\to\infty} u_k$ exist (regardless of the sequence $(a_n)_{n=1}^{\infty}$)? (Recall that we allow ∞ or $-\infty$ as a valid limit.)

6. We write $\liminf a_n = \lim_{k\to\infty} u_k$. Use the formula you recorded in column 2 of the table for question 2 to help you determine $\liminf a_n$. Record your answer in column 3.

7. For the sequence
$$\left(a_n = \frac{2n + 11}{n + 2}\right)_{n=1}^{\infty},$$
find the first 6 terms of the sequence $(u_k)_{k=1}^{\infty}$ and find $\liminf a_n$.

8. For the sequence
$$\left(a_n = \frac{n - 1}{6n}\right)_{n=1}^{\infty},$$
find the first 6 terms of the sequence $(u_k)_{k=1}^{\infty}$ and find $\liminf a_n$.

9. For the sequence

$$\left(a_n = \begin{cases} 2 + \dfrac{1}{n}, & \text{if } n \text{ is even} \\ 0, & \text{if } n \text{ is odd} \end{cases} \right)_{n=1}^{\infty},$$

find the first 6 terms of the sequence $(u_k)_{k=1}^{\infty}$ and find $\liminf a_n$.

8.4 Critical Thinking Questions

1. Use your findings from Sections 8.2 and 8.3 to complete the following table.

sequence $= (a_n)_{n=1}^{\infty}$	$\liminf(a_n)$	$\lim\limits_{n \to \infty} a_n$	$\limsup(a_n)$
$(1/n)_{n=1}^{\infty}$			
$(\cos(n\pi/2))_{n=1}^{\infty}$			
$(2 + ((-1)^n/n))_{n=1}^{\infty}$			
$((n+1)/n)_{n=1}^{\infty}$			
$\left(\begin{cases} 3 - e^{-n}, & \text{if } n \text{ is even} \\ 3, & \text{if } n \text{ is odd} \end{cases} \right)_{n=1}^{\infty}$			

2. Let $(a_n)_{n=1}^{\infty}$ represent an arbitrary sequence. Is one of the following statements always true:

$$\limsup a_n \leq \liminf a_n \quad \text{or} \quad \liminf a_n \leq \limsup a_n?$$

If so, which is true? Can you explain why it is true?

3. Let $(a_n)_{n=1}^{\infty}$ be an arbitrary sequence. Assuming $\lim_{n \to \infty} a_n$ exists, list from smallest to largest the numbers: $\limsup a_n$, $\liminf a_n$, and $\lim_{n \to \infty} a_n$.

4. If the sequence $(a_n)_{n=1}^{\infty}$ converges, what can we say about $\liminf a_n$ and $\limsup a_n$?

5. In all of the examples completed thus far, v_k was a real number. However, consider the sequence $(a_n) = (n)_{n=1}^{\infty}$.

   (a) What is v_1, v_{10}, v_{100}?

   (b) If $v_k = \infty$ for some $k \in \mathbb{N}$, explain why $v_k = \infty$ for all $k \in \mathbb{N}$.

   (c) Note that in the case that $v_k = \infty$ for all $k \in \mathbb{N}$, we define $\limsup a_n = \infty$. For the sequence

   $$(a_n) = \left(\frac{n^2}{n+4} \right)_{n=1}^{\infty},$$

   find $\limsup a_n$.

6. In all of the examples completed thus far, u_k was a real number. However, consider the sequence $(a_n) = (-\exp(n))_{n=1}^{\infty}$.

   (a) What is u_1, u_{10}, u_{100}?

   (b) If $u_k = -\infty$ for some $k \in \mathbb{N}$, explain why $u_k = -\infty$ for all $k \in \mathbb{N}$.

   (c) Note that in the case that $u_k = -\infty$ for all $k \in \mathbb{N}$, we define $\liminf a_n = -\infty$. For the sequence

   $$(a_n) = \left(-\exp(n)\right)_{n=1}^{\infty},$$

   find $\liminf a_n$.

8.5 Questions for Reflection

1. How would we show, using $\limsup a_n$ and $\liminf a_n$, that a sequence does not have a limit?

2. Can you conjecture a condition involving $\limsup a_n$ and $\liminf a_n$ that is equivalent to the convergence of $(a_n)_{n=1}^{\infty}$?

3. Let $(a_n)_{n=1}^{\infty}$ be a sequence. Show that $\liminf a_n = -\limsup -a_n$.

4. Suppose the sequence $(a_n)_{n=1}^{\infty}$ is strictly increasing and $\sup\{a_n : n \in \mathbb{N}\} = B$, where B is a real number. Give a formula for u_k and v_k, where $k \in \mathbb{N}$. What are $\limsup a_n$ and $\liminf a_n$ in this case?

5. Suppose the sequence $(a_n)_{n=1}^{\infty}$ is strictly decreasing and $\inf\{a_n : n \in \mathbb{N}\} = C$, where C is a real number. Give a formula for u_k and v_k where $k \in \mathbb{N}$. What are $\limsup a_n$ and $\liminf a_n$ in this case?

6. Name a condition on the sequence $(a_n)_{n=1}^{\infty}$ that guarantees that the terms u_k and v_k for $k \in \mathbb{N}$ are finite.

7. Let $(a_n)_{n=1}^{\infty}$ be a sequence. We define a *subsequence* of this sequence to be a sequence $(b_k)_{k=1}^{\infty}$ such that for each k, there exists an n_k such that

   $$n_1 < n_2 < n_3 < \cdots < n_k \cdots \qquad \text{and} \qquad b_k = a_{n_k}.$$

   Go back to each sequence considered in this lab, and try to answer the following questions regarding subsequences.

   (a) Is there a case where it is obvious that some subsequence of the sequence has the limit of $\limsup a_n$? If so, what is it?

   (b) Do you think that there is always a subsequence of the sequence for which the limit is $\limsup a_n$? Why or why not?

   (c) Is there a case where it is obvious that some subsequence of the sequence has the limit of $\liminf a_n$? If so, what is it?

   (d) Do you think that there is always a subsequence of the sequence for which the limit is $\liminf a_n$? Why or why not?

9

Continuity and Sequences

9.1 Introduction

In this lab, you are asked to formulate a sequence-based definition of the continuity of a function. Let f denote a function. Given a sequence $(x_n)_{n=1}^{\infty}$ that converges to a domain point $x = x_0$ of f, we are interested in determining the relationship between the continuity of f at $x = x_0$ and the behavior of the corresponding sequence of outputs $(f(x_n))_{n=1}^{\infty}$. This relationship will form the basis of the definition. In the Questions for Reflection, you will learn that this definition is a useful tool in proving continuity theorems involving algebraic combinations of functions.

In Lab 10, you will be introduced to a second, equivalent definition of continuity and then asked to prove that the Lab 9 and Lab 10 definitions are logically equivalent.

Why are we considering two different definitions of continuity? Depending upon the specific problem situation, the two definitions vary in their usefulness. The sequence definition that you derive in this lab is a convenient tool to prove the continuity of a polynomial. The sequence definition is also effective in proving discontinuity. The definition considered in Lab 10 is extremely important when considering a stronger form of continuity called uniform continuity.

In the process of formulating the sequence definition of continuity, you are asked to consider a number of examples in which you apply your intuitive understanding. We review these ideas in the next section. For this lab, you may wish to use the accompanying *Visual Guide for Lab 9*, which can be found at www.saintmarys.edu/~jsnow. Its purpose is to help you visualize the relationship between the graph of a function and the graph of a sequence of outputs of that function. The *Visual Guide* also appears in the appendix. Alternatively, you may use the *Maplet for Lab 9,* which can be found at www.saintmarys.edu/~jsnow.

9.2 Intuitive Notions of Continuity

Your first idea of the continuity of a function is likely linked to its graph. A function f is continuous at a point $x = x_0$ if the graph of f has "no break" at $x = x_0$. Another characterization is that one can draw the graph without picking up one's pencil. In order for this to occur, the function must be defined at $x = x_0$, that is, the value of the real number $f(x_0)$ exists, and, as x approaches x_0, the graph of f approaches $f(x_0)$, that is, $\lim_{x \to x_0} f(x)$ exists and is equal to $f(x_0)$.

From this idea, you were probably introduced to a three-step definition that was stated along the following lines: A function f is *continuous at* $x = x_0$ if and only if

(i) $f(x_0)$ exists;

(ii) $\lim_{x \to x_0} f(x)$ exists; and

(iii) $\lim_{x \to x_0} f(x) = f(x_0)$.

Using this definition, you may have determined the continuity of a given function by analyzing its behavior in three different contexts: graphical, numeric, and algebraic.

For example, consider the function g defined by

$$g(x) = \begin{cases} \dfrac{x^2 - 9}{x - 3}, & \text{if } x \neq 3 \\ 6, & \text{if } x = 3 \end{cases}.$$

Draw the graph of g and visually inspect it near $x = 3$ to determine whether the three points of the definition are satisfied. What can you conclude?

A numerical approach would involve evaluating the function near $x = 3$. An example of possible test points is given in the table below.

x	2.9	2.99	2.999	3.001	3.01	3.1
$g(x)$	5.9	5.99	5.999	6.001	6.01	6.1

Based upon the data presented in the table, it appears that the limit is 6. This suggests that (ii) holds. Does it follow that (i) and (iii) hold? Why, or why not?

Determining the limit of a function using algebraic techniques involves evaluating the limit of the function at the point under consideration. The validity of this approach depends upon a technical definition of limit that is considered in Lab 10.

$$\begin{aligned} \lim_{x \to 3} \frac{x^2 - 9}{x - 3} &= \lim_{x \to 3} \frac{(x + 3)(x - 3)}{x - 3} \\ &= \lim_{x \to 3} (x + 3) \\ &= 3 + 3 \\ &= 6. \end{aligned}$$

As you can see, the $(x - 3)$ term has been canceled. However, if x assumes the value of 3, the resulting 0 in both the numerator and denominator cannot be divided. Fortunately, this is not a problem, because there is an implicit assumption that $x \neq 3$. Yet, in the last step, 3 is substituted for x. This appears to contradict this assumption. Why is this not the case?

Not surprisingly, the numerical and algebraic approaches return the same result. Both establish the existence of the limit at $x = 3$. However, neither approach automatically guarantees (i) and (iii), which must be checked separately.

If we can determine continuity using these different approaches, why is it necessary to devise a different definition? The three-step definition given above, while helpful in understanding the concept of continuity, is not particularly useful in proving theorems. Moreover, it is actually based upon a more rigorous foundation, which is exactly the subject of this and the next lab.

9.3 Using Examples to Enhance Understanding

For each function f_i, $i = 1, 2, 3, 4, 5$, given in the following table, we are interested in describing the relationship between the behavior of f_i near the selected domain point x_0 and a sequence $(f_i(x_n))_{n=1}^{\infty}$, where $(x_n)_{n=1}^{\infty}$ converges to x_0. Complete questions 1–5 as described below for each function f_i, and record your responses in the table.

1. Use numerical, graphical, and/or algebraic methods to determine whether $\lim_{x \to x_0} f_i(x)$ exists. If so, write the value of the limit. If not, write DNE. Enter your response in column Q1.

2. Enter the value of $f_i(x_0)$ in column Q2.

3. For each i, use numerical, graphical, and/or algebraic methods you developed in calculus to determine whether f_i is continuous at $x = x_0$. If f_i is continuous at x_0, write the letter C. If f_i is not continuous at x_0, write NC. Enter your response in column Q3.

4. Verify for yourself that $(a_n)_{n=1}^{\infty}$ converges to x_0. Determine whether the sequence of outputs $\left(f_i(a_n)\right)_{n=1}^{\infty}$ converges; that is, determine whether $\lim_{n \to \infty} f_i(a_n)$ exists. If so, record its value in the table. If not, write DNE. Enter your response in column Q4.

5. Repeat question 4 for the sequence $(b_n)_{n=1}^{\infty}$. Enter your response for this sequence in column Q5.

i	f_i, x_0, Sequences	Q1	Q2	Q3	Q4	Q5
1	$f_1(x) = x^2 - 1$, $\quad x_0 = 0$ $(a_n)_{n=1}^{\infty} = \left(\dfrac{1}{n}\right)_{n=1}^{\infty}$ $(b_n)_{n=1}^{\infty} = \left(\dfrac{n}{n^2 + 1}\right)_{n=1}^{\infty}$					
2	$f_2(x) = \begin{cases} x - 2, & \text{if } x < 4 \\ 2, & \text{if } x = 4 \\ 6 - 2x, & \text{if } x > 4 \end{cases}$, $\quad x_0 = 4$ $(a_n)_{n=1}^{\infty} = \left(4 + \dfrac{(-1)^n}{n}\right)_{n=1}^{\infty}$ $(b_n)_{n=1}^{\infty} = \left(4 - \dfrac{1}{n}\right)_{n=1}^{\infty}$					

i	f_i, x_0, Sequences	Q1	Q2	Q3	Q4	Q5
3	$f_3(x) = \begin{cases} 1/x, & \text{if } x \neq 0 \\ 2, & \text{if } x = 0 \end{cases}$, $\quad x_0 = 0$ $(a_n)_{n=1}^{\infty} = \left(-\dfrac{1}{n} \right)_{n=1}^{\infty}$ $(b_n)_{n=1}^{\infty} = \left(\dfrac{1}{n^2} \right)_{n=1}^{\infty}$					
4	$f_4(x) = \begin{cases} \dfrac{x^2 - 2x - 15}{x - 5}, & \text{if } x \neq 5 \\ 1, & \text{if } x = 5 \end{cases}$, $\quad x_0 = 5$ $(a_n)_{n=1}^{\infty} = \left(\dfrac{5n}{n+1} \right)_{n=1}^{\infty}$ $(b_n)_{n=1}^{\infty} = \left(5 + \dfrac{(-1)^n}{n} \right)_{n=1}^{\infty}$					
5	$f_5(x) = \begin{cases} 7 - x, & \text{if } x < 2 \\ 2x + 1, & \text{if } x \geq 2 \end{cases}$, $\quad x_0 = 2$ $(a_n)_{n=1}^{\infty} = \left(2 - \dfrac{1}{n} \right)_{n=1}^{\infty}$ $(b_n)_{n=1}^{\infty} = \left(\dfrac{2n}{n+3} \right)_{n=1}^{\infty}$					

9.4   Critical Thinking Questions

Use the data gathered in the previous section to answer the following questions.

1. Consider those i for which f_i is continuous at $x = x_0$.

   (a) For the given sequences $(a_n)_{n=1}^{\infty}$ and $(b_n)_{n=1}^{\infty}$, do both sequences $(f_i(a_n))_{n=1}^{\infty}$ and $(f_i(b_n))_{n=1}^{\infty}$ converge? If so, are their limits equal? Is this true for every i for which f_i is continuous?

   (b) Let f be a function that is continuous at $x = x_0$. Use the data you have gathered to describe the relationship, if any, between

   $$\lim_{x \to x_0} f(x), \qquad \lim_{n \to \infty} f(a_n), \quad \text{and} \quad \lim_{n \to \infty} f(b_n),$$

   where $(a_n)_{n=1}^{\infty}$ and $(b_n)_{n=1}^{\infty}$ are both sequences that converge to x_0. Specifically, if you claim that the limits exist and are equal, try to explain why this is the case. If you claim that the limits do not necessarily have to be equal or do not necessarily have to exist, try to explain why this occurs.

   (c) Let f be a function that is continuous at $x = x_0$. Use the data you have gathered to describe the relationship, if any, between

$$f(x_0), \qquad \lim_{n \to \infty} f(a_n), \qquad \text{and} \qquad \lim_{n \to \infty} f(b_n),$$

where $(a_n)_{n=1}^\infty$ and $(b_n)_{n=1}^\infty$ are both sequences that converge to x_0. Specifically, if you claim that the limits exist and are equal to $f(x_0)$, try to explain why this is the case. If you claim that the limits do not necessarily have to be equal to $f(x_0)$, try to explain why this occurs.

2. Consider those i for which f_i is discontinuous at $x = x_0$.

(a) For the given sequences $(a_n)_{n=1}^\infty$ and $(b_n)_{n=1}^\infty$, do both sequences $(f_i(a_n))_{n=1}^\infty$ and $(f_i(b_n))_{n=1}^\infty$ converge? If so, are their limits equal? If there is an i for which

$$\lim_{n \to \infty} f_i(a_n) = \lim_{n \to \infty} f_i(b_n),$$

could you define another sequence $(c_n)_{n=1}^\infty$ that converges to x_0 such that

$$\lim_{n \to \infty} f_i(c_n) \neq \lim_{n \to \infty} f_i(a_n)?$$

(b) Let f be a function that is discontinuous at $x = x_0$. Use the data you have gathered to describe the relationship, if any, between

$$\lim_{x \to x_0} f(x), \qquad \lim_{n \to \infty} f(a_n), \qquad \text{and} \qquad \lim_{n \to \infty} f(b_n),$$

where $(a_n)_{n=1}^\infty$ and $(b_n)_{n=1}^\infty$ are both sequences that converge to x_0. Specifically, if you claim that the limits exist and are equal, try to explain why this is the case. If you claim that the limits do not necessarily have to be equal or do not necessarily have to exist, try to explain why this occurs.

(c) Let f be a function that is discontinuous at $x = x_0$. Use the data you have gathered to describe the relationship, if any, between

$$f(x_0), \qquad \lim_{n \to \infty} f(a_n), \qquad \text{and} \qquad \lim_{n \to \infty} f(b_n),$$

where $(a_n)_{n=1}^\infty$ and $(b_n)_{n=1}^\infty$ are both sequences that converge to x_0. Specifically, if you claim that the limits exist and are equal to $f(x_0)$, try to explain why this is the case. If you claim that the limits do not necessarily have to be equal to $f(x_0)$, try to explain why this occurs.

On the basis of your answers to items 1 and 2, answer the following more general questions.

3. If a function f is discontinuous at $x = x_0$, is it possible to define a sequence $(x_n)_{n=1}^\infty$ that converges to x_0 such that the sequence of outputs $(f(x_n))_{n=1}^\infty$ converges to $f(x_0)$? If so, indicate an i for which this occurs. If the answer to your question is no, try to explain why you can't find such a sequence.

4. If a function f is continuous at $x = x_0$, does it seem possible to define a sequence $(x_n)_{n=1}^\infty$ that converges to x_0 such that the sequence of outputs $(f(x_n))_{n=1}^\infty$ does not converge to $f(x_0)$? If so, indicate an i for which this occurs. If the answer to your question is no, try to explain why you can't find such a sequence.

5. In the table that follows, you are given five functions, g_1, g_2, g_3, g_4, and g_5, for which the only information provided is that given in the table. Use the given information, together with what you have learned from the examples in Section 9.3, to fill in each missing blank.

g_i	$g_i(x_0)$	Behavior of $(g_i(x_n))_{n=1}^{\infty}$, where $(x_n)_{n=1}^{\infty}$ converges to x_0	Continuous at x_0?
g_1	2	There exists $(x_n) \longrightarrow x_0$ such that $\lim\limits_{n \to \infty} g_1(x_n)$ DNE.	
g_2	DNE	For all $(x_n) \longrightarrow x_0$, $\lim\limits_{n \to \infty} g_2(x_n) = 3$.	
g_3	3		C
g_4	-2		NC
g_5		For all $(x_n) \longrightarrow x_0$, $\lim\limits_{n \to \infty} g_5(x_n) = -1$.	C

6. Consider again the five example functions from Section 9.3. In a similar manner, fill in the following table for these five functions.

f_i	$f_i(x_0)$	Behavior of $(f_i(x_n))_{n=1}^{\infty}$, where $(x_n)_{n=1}^{\infty}$ converges to x_0	Continuous at x_0?
f_1			
f_2			
f_3			
f_4			
f_5			

7. Based on your findings and the exercises you have just completed, try to formulate a definition of continuity in terms of the behavior of sequences. Specifically, what is the relationship between the continuity of a function f at a point $x = x_0$ and the behavior of a sequence $(f(x_n))_{n=1}^{\infty}$ for which the sequence $(x_n)_{n=1}^{\infty}$ converges to x_0?

9.5   Questions for Reflection

1. Suppose f is a function. Suppose there exists at least one sequence $(c_n)_{n=1}^{\infty}$ that converges to x_0 such that $\lim_{n \to \infty} f(c_n) = -1$. Assume there is at least one other sequence $(d_n)_{n=1}^{\infty}$ that converges to x_0 such that $\lim_{n \to \infty} f(d_n) = -2$. Suppose that $f(x_0) = -2$. Can we conclude that g is continuous at $x = x_0$? Discontinuous? Or, is there insufficient information to make a determination? Explain your answer.

2. Suppose g is a function such that $\lim_{n \to \infty} g(x_n) = 2$ for every sequence $(x_n)_{n=1}^{\infty}$ that converges to x_0. Can we conclude that g is continuous at $x = x_0$? Discontinuous? Or, is there insufficient information to make a determination? Explain your answer.

3. Suppose that h is a function such that $\lim_{n \to \infty} h(x_n)$ does not exist for some sequence $(x_n)_{n=1}^{\infty}$ that converges to x_0. Can we conclude that h is continuous at $x = x_0$? Discontinuous? Or, is there insufficient information to make a determination? Explain your answer.

4. Suppose that r is a function such that $r(x_0) = 3$ and such that $\lim_{n \to \infty} r(x_n) = 3$ for every sequence $(x_n)_{n=1}^{\infty}$ that converges to x_0. Can we conclude that r is continuous at $x = x_0$? Discontinuous? Or, is there insufficient information to make a determination? Explain your answer.

5. Suppose that t is a function such that $t(x_0) = 2$ and such that $\lim_{n \to \infty} t(x_n) = -4$ for every sequence $(x_n)_{n=1}^{\infty}$ that converges to x_0. Can we conclude that t is continuous at $x = x_0$? Discontinuous? Or, is there insufficient information to make a determination? Explain your answer.

6. Let h be a function, and let $c \in \mathbf{R}$ be a constant. If h is continuous at x_0, use the sequence definition of continuity to show that the function ch is continuous at $x = x_0$.

7. If f and g are continuous at $x = x_0$, use the sequence definition of continuity to show that $(f + g)$ is continuous at $x = x_0$.

8. If f and g are continuous at $x = x_0$, use the sequence definition of continuity to show that $(f \cdot g)$ is continuous at $x = x_0$.

9. If f and g are continuous at $x = x_0$, and if $g(x_0) \neq 0$, use the sequence definition of continuity to show that (f/g) is continuous at $x = x_0$.

10. Let f be a function defined by

$$f(x) = a_0 + a_1 x + a_2 x^2 + \cdots + a_n x^n.$$

Use any results formulated or proven in this lab to show that f is continuous at every point $x \in \mathbf{R}$.

11. A rational function is defined by an expression that is the quotient of two polynomials. In other words, r is a rational function if there exist two polynomial functions p and q such that

$$r(x) = \frac{p(x)}{q(x)}.$$

What is the domain of r? At what points will r be continuous? State and prove a theorem that describes where r is continuous.

12. To what degree would the sequence definition of continuity need to be modified in order to be suitable as a definition for the limit of a function? In other words, if f is a function, and if $(x_n)_{n=1}^{\infty}$ is any sequence of domain points such that $(x_n)_{n=1}^{\infty}$ converges to x_0, then $\lim_{x \to x_0} f(x) = L$ if and only if ...?

10

![chapter number banner]

Another Definition of Continuity

10.1 Introduction

In Lab 9, you formulated a definition of continuity in terms of the behavior of sequences. In this lab, we consider a second definition, often referred to as the "$\epsilon - \delta$" definition of continuity. Although logically equivalent to the sequence-based definition, it differs in the sense that it involves relationships between intervals of the domain and range. Rather than have you derive the definition, this lab is designed to help you grasp how it works. Its statement, which is given below and which may be familiar to you from calculus, includes two quantifiers and a functional relationship between those quantifiers that often make the definition difficult to understand.

Definition. Let $y = f(x)$ be a function. Let $x = x_0$ be a point of the domain of f. The function f is said to be *continuous* at $x = x_0$ if and only if given $\epsilon > 0$, there exists $\delta > 0$ such that if $x \in (x_0 - \delta, x_0 + \delta)$, then $f(x) \in (f(x_0) - \epsilon, f(x_0) + \epsilon)$.

In Section 10.5, Questions for Reflection, you are asked to prove that the $\epsilon - \delta$ definition is logically equivalent to the sequence definition you derived in Lab 9.

With some fairly minor, although important modifications, the $\epsilon - \delta$ definition of continuity can be restructured to serve as the $\epsilon - \delta$ definition of the limit of a function. This is the focus of Section 10.4. In this section, you will be able to see how the two concepts of limit and continuity are related, while also considering their important differences. An additional purpose of this lab, as in the cases of earlier labs like Lab 3 and Lab 9, is to make precise the notions of continuity and limit, so that you can use them to prove theorems in real analysis. For several of the activities, you may wish to use the accompanying *Visual Guide for Lab 10*, which can be obtained at www.saintmarys.edu/~jsnow. The *Visual Guide* also appears in the appendix. Alternatively, you may use the *Maplet for Lab 10,* which can be found at www.saintmarys.edu/~jsnow.

10.2 Using Examples to Enhance Understanding

For each function f_i, $i = 1, 2, 3, 4, 5, 6, 7$, complete questions 1 and 2, and record your findings in the designated columns of the following table.

$f_i,\ x_0$	$f(x_0)$	C/NC	$\epsilon = 1$	$\epsilon = .5$	$\epsilon = .1$	$\epsilon = .01$
$f_1(x) = |x|,\quad x_0 = 0$						
$f_2(x) = \begin{cases} \dfrac{x^2 - 16}{x - 4}, & \text{if } x \neq 4 \\ 31/4, & \text{if } x = 4 \end{cases},\quad x_0 = 4$						
$f_3(x) = \begin{cases} x^2, & \text{if } x < 1 \\ 1, & \text{if } x = 1, \\ 2x - 49/50, & \text{if } x > 1 \end{cases} \ x_0 = 1$						
$f_4(x) = \begin{cases} \sin\left(\frac{1}{x}\right), & \text{if } x \neq 0 \\ 0, & \text{if } x = 0 \end{cases},\quad x_0 = 0$						
$f_5(x) = \begin{cases} x\sin\left(\frac{1}{x}\right), & \text{if } x \neq 0 \\ 0, & \text{if } x = 0 \end{cases},\quad x_0 = 0$						
$f_6(x) = \begin{cases} \frac{1}{x}, & \text{if } x \neq 0 \\ 2, & \text{if } x = 0 \end{cases},\quad x_0 = 0$						
$f_7(x) = \begin{cases} 0, & \text{if } x \in \mathbf{Q} \\ 1, & \text{if } x \notin \mathbf{Q} \end{cases},\quad x_0 = 0$						

1. In column 2, enter the value of $f_i(x_0)$. In column 3, write C, if the function is continuous at x_0, or NC, if the function is discontinuous at x_0. Use graphical, numerical, or algebraic methods that you developed in calculus or use the sequence definition you derived in Lab 9 to help you make these determinations.

2. Your responses in columns 4–7 are based upon constructing an ϵ-band around $f_i(x_0)$. By an ϵ-band around $f_i(x_0)$, we mean the interval $(f_i(x_0) - \epsilon, f_i(x_0) + \epsilon)$, an open interval of length 2ϵ along the y-axis in which the point $f_i(x_0)$ is situated at the center of the interval. For example, in the case of the function f_1, the ϵ-band of radius $\epsilon = 1$ about the point $x_0 = 0$ is given by $(f_1(0) - 1, f_1(0) + 1)$ and illustrated in the following figure.

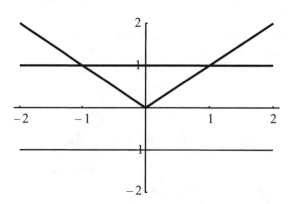

Once you have such a graph, your task, for each i and for each ϵ given in the table, is to determine whether you can find a positive number δ such that

$$\text{if } x \in (x_0 - \delta, x_0 + \delta), \text{ then } f_i(x) \in (f_i(x_0) - \epsilon, f_i(x_0) + \epsilon).$$

If you can find such a δ for the indicated value of ϵ, list it in the appropriate cell of the table. If you cannot, write No δ. As you work through these examples, think about the relationship between ϵ and δ as given in the statement of the definition presented in the introduction.

10.3 Critical Thinking Questions

Use the data in the table to answer the following questions.

1. For those i for which you were able to find a δ for $\epsilon = .1$, select three different x that fall in the interval $(x_0 - \delta, x_0 + \delta)$, and show that $f_i(x) \in (f_i(x_0) - \epsilon, f_i(x_0) + \epsilon)$.

2. In those cases in which you were able to find a δ, was the value of δ unique? Could you have selected another value for δ? Could you have selected a maximum value for δ? Could you have selected a minimum value for δ?

3. In those cases in which you could not find a δ for a particular ϵ, can you explain what went wrong? In each such case, identify those x in $(x_0 - \delta, x_0 + \delta)$ for which $f_i(x) \notin (f_i(x_0) - \epsilon, f_i(x_0) + \epsilon)$.

4. You considered only four possibilities for ϵ. For those f_i for which f_i is continuous at $x = x_0$, why must we be able to find a δ for any $\epsilon > 0$, large or small?

5. For those f_i that are discontinuous at $x = x_0$, would the answer to the question, "Can we find a δ?", be the same for every $\epsilon > 0$, large or small? Provide a thorough explanation.

6. For those functions f_i that are discontinuous at $x = x_0$, determine which of the following best summarizes your findings and understanding.

    (a) If f_i is discontinuous at $x = x_0$, and if $f_i(x_0)$ exists, then, for every $\epsilon > 0$, we can find a $\delta > 0$ such that $f_i(x) \notin (f_i(x_0) - \epsilon, f_i(x_0) + \epsilon)$ for at least one $x \in (x_0 - \delta, x_0 + \delta)$.

    (b) If f_i is discontinuous at $x = x_0$, and if $f_i(x_0)$ exists, then there exists $\delta > 0$ such that for every $\epsilon > 0$, $f_i(x) \notin (f_i(x_0) - \epsilon, f_i(x_0) + \epsilon)$ for at least one $x \in (x_0 - \delta, x_0 + \delta)$.

    (c) If f_i is discontinuous at $x = x_0$, then there is an $\epsilon > 0$ such that for every $\delta > 0$, $f_i(x) \notin (f_i(x_0) - \epsilon, f_i(x_0) + \epsilon)$ for at least one $x \in (x_0 - \delta, x_0 + \delta)$.

7. Describe the relationship between ϵ and δ. Does it appear as though δ is a function of ϵ, or ϵ is a function of δ? Which of these acts as the independent variable? Which is the dependent variable? Why?

8. Suppose it were possible to write a computer program based upon the $\epsilon - \delta$ definition. Since a computer can only execute a finite number of steps, suppose that there are only four possible values for ϵ, $\epsilon = .001, .01, .1, 1$, and that there are only five possible values for δ, $\delta = 1, .5, .05, .005, .0005$. If the program is designed to accept a function f and a domain point x_0 and to return either "the function f is continuous at $x = x_0$," or "the function f is not continuous at $x = x_0$," explain how the program works. If you were in fact to write and

then to execute such a program, which of the two incorrect responses is more likely to occur when a function f and a point x_0 are entered: that f is discontinuous at x_0 when in fact it is continuous, or that f is discontinuous at x_0 when it is actually continuous at x_0?

10.4   Definition of the Limit of a Function

In an informal sense, a function f has a limit L at x_0 if, as x gets "closer and closer" to x_0, the corresponding output values $f(x)$ get "closer and closer" to L. The definition of the limit of a function at a point is similar to the definition of continuity. In order to see the differences, which, although subtle, are crucial, consider the following examples given in the table. Enter your results for questions 1 and 2 in columns 1 and 2 of the table. Your responses to question 3 are to be entered in columns 3 and 4.

h_i, x_0	L	$L = h_i(x_0)$?	$\epsilon = .5$	$\epsilon = .1$
$h_1(x) = \dfrac{x^2 - 9}{x - 3}, \quad x_0 = 3$				
$h_2(x) = \begin{cases} 3 - x, & \text{if } x < 1 \\ 1, & \text{if } x = 1, \\ 3x - 1, & \text{if } x > 1 \end{cases} \quad x_0 = 1$				
$h_3(x) = \dfrac{x - 4}{\sqrt{x} - 2}, \quad x_0 = 4$				
$h_4(x) = x \sin\left(\frac{1}{x}\right), \quad x_0 = 0$				

1. Using graphical, numeric, and/or algebraic techniques you developed in calculus (and reviewed in Lab 9), identify the limit of h_i, $i = 1, 2, 3, 4$, at $x = x_0$, and record your answer in column 2.

2. Does $h_i(x_0)$, $i = 1, 2, 3, 4$, exist? If so, is $h_i(x_0) = L$? If not, explain why this does not affect the existence of the limit.

3. For each $\epsilon = .5, .1$, construct an ϵ-band

$$(L - \epsilon, L + \epsilon)$$

around the limit L you identified in question 1. Describe which input values x satisfy

$$h_i(x) \in (L - \epsilon, L + \epsilon),$$

and enter your responses in the appropriate column.

4. If an appropriate restriction were made, could you construct a suitable δ-band for each value of ϵ given in the table? If so, state the restriction.

5. If you selected any value of $\epsilon > 0$, large or small, and assumed the restriction you made in question 4, would you always be able to find a corresponding δ? Explain your answer.

6. Use the information you have gathered to formulate a definition for the limit of a function at a point.

7. In what way is the $\epsilon - \delta$ definition of limit that you have just come up with consistent with the informal notion of the limit mentioned at the beginning of this section?

10.5 Questions for Reflection

1. Show that the sequence definition of continuity you formulated in Lab 9 is logically equivalent to the $\epsilon - \delta$ definition of continuity. Specifically, show that the sequence definition holds if and only if the $\epsilon - \delta$ definition holds.

2. For each of the following statements, identify the error, and explain how the $\epsilon - \delta$ definition of continuity is violated. (Your explanation may be based upon presenting a suitable example that illustrates the error.)

   (a) Let f be a function, and suppose that $x_0 = 2$. For $\epsilon > .2$, there always exists a δ such that if $x \in (2 - \delta, 2 + \delta)$, then $f(x) \in (f(2) - \epsilon, f(2) + \epsilon)$. Therefore, f is continuous at $x = 2$.

   (b) Let g be a function, and suppose that $x_0 = -3$. There exists an $\epsilon > 0$ such that for every $\delta > 0$ it follows that if $x \in (-3 - \delta, -3 + \delta)$, then $g(x) \in (g(-3) - \epsilon, g(-3) + \epsilon)$. Therefore, g is continuous at $x = -3$.

   (c) Let r be a function, and suppose that $x_0 = -2$. For every $\delta > 0$, we can find an $\epsilon > 0$ such that if $x \in (-2 - \delta, -2 + \delta)$, then $r(x) \in (r(-2) - \epsilon, r(-2) + \epsilon)$. Therefore, r is continuous at $x_0 = -2$.

3. Formulate the definitions of one-sided limits. Although you may have considered one-sided limits in calculus, let's take a few moments to think about this intuitively. Exactly what do we mean by a one-sided limit? Let f be a function, and let $x = x_0$ be a point in the domain of f. If f has a left-sided limit L at x_0, then, as x gets nearer x_0 from the left (where $x < x_0$), the corresponding functional values $f(x)$ approach L, but as x gets nearer x_0 from the right (where $x > x_0$), the corresponding functional values may not approach L. Your task in part (a) is to express this formally. In part (b), you are asked to write the definition of a right-sided limit.

   (a) If the limit "from the left" exists, that is, $\lim_{x \to x_0^-} f(x) = L$, but the limit "from the right" does not exist or does not equal L, what modification do we need to make to the definition of the limit of a function you formulated in Section 10.4 to come up with the "left-side" definition of the limit of f at $x = x_0$? In other words, for a function f, $\lim_{x \to x_0^-} f(x) = L$ if and only if ...?

   (b) Similarly, if the "limit from the right" exists, that is $\lim_{x \to x_0^+} f(x) = L$, but the limit "from the left" does not exist or does not equal L, in what way do we need to modify the definition of the limit of a function to come up with the "right-side" definition of limit? In other words, for a function f, $\lim_{x \to x_0^+} f(x) = L$ if and only if ...?

4. State the negation of the $\epsilon - \delta$ definition of continuity. Use this statement to prove that a function f defined by

$$f(x) = \begin{cases} \dfrac{|x|}{x}, & \text{if } x \neq 0 \\ 0, & \text{if } x = 0 \end{cases}$$

is discontinuous at $x = 0$?

5. For each situation described below, give an example of a discontinuous function that satisfies the given conditions:

(a) A function f for which there exists an $\epsilon > 0$ such that for every $\delta > 0$,

$$f(x) \in (f(x_0) - \epsilon, f(x_0) + \epsilon)$$

whenever $x \in (x_0 - \delta, x_0 + \delta)$.

(b) A function g such that for every $\delta > 0$ there exists an $\epsilon > 0$ such that

$$f(x) \in (f(x_0) - \epsilon, f(x_0) + \epsilon)$$

whenever $x \in (x_0 - \delta, x_0 + \delta)$.

(c) A function h such that for every $\epsilon > 0$ there is no $\delta > 0$ such that

$$f(x) \in (f(x_0) - \epsilon, f(x_0) + \epsilon)$$

whenever $x \in (x_0 - \delta, x_0 + \delta)$.

6. Suppose you were asked to tutor a freshman calculus student who had just seen the $\epsilon - \delta$ definition of limit for the first time. How would you explain the definition to this student in a way that would help her or him to understand how it works? You may use a picture to assist with your explanation.

11

Experience with the $\epsilon - \delta$ Definitions of Continuity and Limit

11.1 Introduction

This lab is a continuation of Lab 10. In that lab, the purpose was to understand how the $\epsilon - \delta$ definitions of continuity and limit work. In this lab, the goal is to learn how to use these definitions to construct proofs. Our work encompasses two different contexts. First, we show how the definition of limit applies to specific functions. Then, we use the definition of limit to prove results involving algebraic combinations of functions. Before continuing, we discuss the issue of expressing the two definitions in absolute value notation rather than interval notation. In Lab 10, you used interval versions of the two definitions:

A function f is *continuous* at a domain point $x = x_0$ if and only if given $\epsilon > 0$, there exists $\delta > 0$ such that if $x \in (x_0 - \delta, x_0 + \delta)$, then $f(x) \in (f(x_0) - \epsilon, f(x_0) + \epsilon)$.

You probably stated the $\epsilon - \delta$ of limit in a similar manner:

The *limit* of a function f at a domain point $x = x_0$ is L if and only if given $\epsilon > 0$, there exists $\delta > 0$ such that if $x \in (x_0 - \delta, x_0 + \delta)$, where $x \neq x_0$, then $f(x) \in (L - \epsilon, L + \epsilon)$.

When studying graphs of functions, as you did in Lab 10, the use of interval notation proves to be quite helpful. However, the process of proving the propositions and theorems in this lab, as well as in other settings in real analysis, is facilitated by using absolute value notation. Using this notation, the definition of continuity can be restated in the following way:

A function f is continuous at a domain point $x = x_0$ if and only if given $\epsilon > 0$, there exists $\delta > 0$ such that if $|x - x_0| < \delta$, then $|f(x) - f(x_0)| < \epsilon$.

Before continuing, write the limit definition using absolute value notation.

Once you have completed this task, you should be ready to write some proofs. In Section 11.2, you are given the same four examples you worked with in Section 10.4 of Lab 10. Rather than trying to formulate the definition as you did there, here you apply the definition to find a formula

67

for δ in terms of ϵ. You then use the formula to prove the existence of the limit. In Section 11.3, the focus is more general, as you use the $\epsilon - \delta$ definition to prove that the limit of an algebraic combination of two functions (with suitable restrictions) is equal to the algebraic combination of the limits. In each case, you are guided toward the general proof by considering an example. After each proof, you are asked to prove continuity, once necessary modifications and/or assumptions have been made. You may recall that in the Lab 9 Questions for Reflection you used the sequence definition of continuity to prove that the sum, product, and quotient (with certain restrictions) of two continuous functions are continuous; the idea here is similar.

11.2 The Limit Definition with Particular Functions

In this section, we examine an algebraic method for finding an expression for δ in terms of ϵ. We use two examples to explain the process.

11.2.1 Example 1

Define a function f by $f(x) = 2x - 11$. We wish to show that $\lim_{x \to 4} f(x) = -3$.

Expression for δ in Terms of ϵ. Let $\epsilon > 0$. Our goal is to find $\delta > 0$ such that

$$\text{if} \quad 0 < |x - 4| < \delta, \quad \text{then} \quad |f(x) - (-3)| < \epsilon.$$

In order to find an expression for δ in terms of ϵ, we work with the expression $|(2x-11)-(-3)| < \epsilon$ and solve for $|x - 4|$:

$$\epsilon > |(2x - 11) + 3|$$
$$= |2x - 8|$$
$$= |2(x - 4)|$$
$$= 2|x - 4|.$$

Dividing both sides of the inequality yields the desired result, $|x - 4| < \epsilon/2$. As a result, we choose $\delta = \frac{\epsilon}{2}$.

Formal Statement and Proof. Let f be a function defined by $f(x) = 2x - 11$. Then, $\lim_{x \to 4} f(x) = -3$.

Proof. Let $\epsilon > 0$. Let $\delta = \dfrac{\epsilon}{2}$. Let x be a real number such that $0 < |x - 4| < \delta$. Then,

$$|f(x) - (-3)| = |(2x - 11) + 3|$$
$$= |2x - 8|$$
$$= |2(x - 4)|$$
$$= 2|x - 4|$$
$$< 2\left(\frac{\epsilon}{2}\right)$$
$$= \epsilon. \qquad \qquad \text{Q.E.D.}$$

11.2.2 Example 2

Sometimes it is difficult or messy to solve for $|x - x_0|$. For example, consider the function g defined by $g(x) = \dfrac{2x + 5}{x - 2}$. We wish to prove that $\lim_{x \to 4} g(x) = \dfrac{13}{2}$.

Expression for δ in Terms of ϵ.   Let $\epsilon > 0$. As before, our goal is to find $\delta > 0$ such that

$$\text{if } 0 < |x - 4| < \delta, \text{ then } \left| \frac{2x + 5}{x - 2} - \frac{13}{2} \right| < \epsilon.$$

Rewriting this inequality, we obtain

$$
\begin{aligned}
\epsilon &> \left| \frac{2x + 5}{x - 2} - \frac{13}{2} \right| \\
&= \left| \frac{-9x + 36}{2(x - 2)} \right| \\
&= \left| \frac{-9(x - 4)}{2(x - 2)} \right| \\
&= \frac{9}{2} \frac{|x - 4|}{|x - 2|}.
\end{aligned}
$$

Dividing both sides of the inequality by $9/2$ yields

$$\frac{|x - 4|}{|x - 2|} < \frac{2}{9}\epsilon.$$

Unlike Example 1, the presence of the $|x - 2|$ term in the denominator prevents us from solving for $|x - 4|$ directly. In order to get an expression in terms of ϵ and $|x - 4|$ exclusively, we require that $\delta < 1$. Using this restriction, we can place a bound on $|x - 2|$ that will allow us to isolate the term $|x - 4|$:

$$|x - 4| < \delta \implies -1 < x - 4 < 1 \implies 1 < x - 2 < 3$$

$$\implies 1 > \frac{1}{x - 2} > \frac{1}{3} \implies \left| \frac{1}{x - 2} \right| < 1.$$

This last inequality allows us to write

$$\left| \frac{-9(x - 4)}{2(x - 2)} \right| = \frac{9}{2} \frac{|x - 4|}{|x - 2|} < \frac{9}{2}|x - 4|, \quad \text{provided } \delta < 1.$$

If we require

$$\frac{9}{2}|x - 4| < \epsilon,$$

then we want $|x - 4|$ to be less than both 1 and $\frac{9}{2}\epsilon$. Hence, we let $\delta = \min\left\{ 1, \frac{2}{9}\epsilon \right\}$.

Formal Statement and Proof.   Let g be a function defined by $g(x) = (2x + 5)/(x - 2)$. Then, $\lim_{x \to 4} g(x) = \frac{13}{2}$.

Proof.   Let $\epsilon > 0$. Let $\delta = \min\left\{1, \frac{2}{9}\epsilon\right\}$. Let x be a real number such that $0 < |x - 4| < \delta$. Then,

$$\left| g(x) - \frac{13}{2} \right| = \left| \frac{2x + 5}{x - 2} - \frac{13}{2} \right|$$

$$= \left| \frac{-9x + 36}{2(x - 2)} \right|$$

$$= \left| \frac{-9(x - 4)}{2(x - 2)} \right|$$

$$= \frac{9}{2} \frac{|x - 4|}{|x - 2|}$$

$$< \frac{9}{2} |x - 4|, \text{ because } \delta < 1$$

$$< \frac{9}{2} \left(\frac{2}{9} \epsilon \right)$$

$$= \epsilon. \hspace{3cm} \text{Q.E.D.}$$

11.2.3   Practice with the Definition

For each function h_i, $i = 1, 2, 3, 4$, enter the value of the limit in column 2. Let $\epsilon > 0$ represent an arbitrary positive real number. Using techniques similar to those given in Examples 1 and 2 in Sections 11.2.1 and 11.2.2, find a formula for δ in terms ϵ. Write your formula for δ in column 3.

1. Use the formula you entered in column 3 to find a specific δ that corresponds to $\epsilon = .1$. Enter this value in column 4.

2. In column 5, enter the value of δ that you recorded for $\epsilon = .1$ in Section 10.4 of Lab 10. Is the definition of limit violated if the value entered in column 5 differs from that entered in column 4? Why, or why not?

h_i	L	δ	$\epsilon = .1$	Lab 10
$h_1(x) = \dfrac{x^2 - 9}{x - 3}, \quad x_0 = 3$				
$h_2(x) = \begin{cases} 3 - x, & \text{if } x < 1 \\ 1, & \text{if } x = 1, \\ 3x - 1, & \text{if } x > 1 \end{cases} \quad x_0 = 1$				
$h_3(x) = \dfrac{x - 4}{\sqrt{x} - 2}, \quad x_0 = 4$				
$h_4(x) = x \sin\left(\dfrac{1}{x}\right), \quad x_0 = 0$				

3. Use the expression you entered in column 3 to write a formal proof in a manner that is similar to the examples considered in Sections 11.2.1 and 11.2.2.

4. The functions h_1, h_2, h_3, and h_4, if suitably redefined at $x = x_0$, can be made to be continuous at $x = x_0$. Redefine these functions so that they are continuous, and then use the $\epsilon - \delta$ definition of continuity, together with the formulas for δ that you have devised, to prove the continuity of each function at $x = x_0$. Why can the formulas for δ, at least in these cases, remain unchanged?

11.3 Algebraic Combinations

In the last section, you used the definition of limit to write proofs involving specific functions. In this section, we prove theorems involving algebraic combinations of functions.

11.3.1 The Sum of Two Functions

Let f and g be the functions defined in Section 11.2. Both functions have limits as $x \to 4$. We show that

$$\lim_{x \to 4} (f + g)(x) = \lim_{x \to 4} f(x) + \lim_{x \to 4} g(x) = -3 + \frac{13}{2}.$$

Discussion. Let $\epsilon > 0$. We want to find $\delta > 0$ such that

$$\text{if} \quad 0 < |x - 4| < \delta, \quad \text{then} \quad \left| \left((2x - 11) + \frac{2x + 5}{x - 2} \right) - \left(-3 + \frac{13}{2} \right) \right| < \epsilon.$$

In Lab 6, in trying to prove the convergence of the sum of two convergent sequences, you may recall that we isolated the terms corresponding to the summands before applying the definition. We do essentially the same thing here: we isolate the terms $|(2x - 11) + 3|$ and $\left| \frac{2x+5}{x-2} - \frac{13}{2} \right|$ and then apply the limit definition to each summand individually:

$$\left| \left((2x - 11) + \frac{2x + 5}{x - 2} \right) - \left(-3 + \frac{13}{2} \right) \right| = \left| [(2x - 11) + 3] + \left(\frac{2x + 5}{x - 2} - \frac{13}{2} \right) \right|$$

$$\leq |(2x - 11) + 3| + \left| \frac{2x + 5}{x - 2} - \frac{13}{2} \right|. \qquad (*)$$

Since we would like the entire sum to be less than ϵ, we decide to require each term in $(*)$ to be less than $\epsilon/2$. Since the limit of each individual function exists at $x = 4$, we can, given $\epsilon/2$, find a corresponding $\delta > 0$ for both individual summands.

Specifically, since $\lim_{x \to 4} (2x - 11) = -3$, we can find $\delta_1 > 0$ such that

$$\text{if} \quad 0 < |x - 4| < \delta_1, \quad \text{then} \quad |(2x - 11) + 3| < \frac{\epsilon}{2}.$$

Go back to Section 11.2.1 and see what modifications, if any, need to be made to find a suitable δ_1. Similarly, since

$$\lim_{x \to 4} \frac{2x + 5}{x - 2} = \frac{13}{2},$$

we can find $\delta_2 > 0$ such that

$$\text{if} \quad 0 < |x - 4| < \delta_2, \quad \text{then} \quad \left| \frac{2x + 5}{x - 2} - 13/2 \right| < \frac{\epsilon}{2}.$$

Use your results from Section 11.2.2 to help in finding a suitable δ_2.

Proof. Let $\epsilon > 0$. Let $\delta =$? (*you are asked to determine this*). Choose x such that $0 < |x-4| < \delta$. Then,

$$\left| \left((2x - 11) + \frac{2x + 5}{x - 2} \right) - (-3 + 13/2) \right| < \epsilon.$$

Q.E.D.

Exercises

1. Complete the proof just outlined by giving the exact form for δ and filling in the missing details of the proof.

2. Suppose h and k are two functions such that $\lim_{x \to x_0} h(x) = L$ and $\lim_{x \to x_0} k(x) = M$. Use the $\epsilon - \delta$ definition of limit, together with the techniques displayed in the example above, to prove that $\lim_{x \to x_0} (h + k)(x) = L + M$.

3. If h and k are both continuous at $x = x_0$, what changes would you have to make in the proof you wrote in the previous exercise to prove the continuity of $h + k$ at $x = x_0$? In Lab 9, you used the sequence definition to prove that the sum of two continuous functions is also continuous. Write a similar proof using the $\epsilon - \delta$ definition of continuity.

11.3.2   The Product of Two Functions

Consider again the functions f and g defined in Section 11.2. We use the $\epsilon - \delta$ definition to show that

$$\lim_{x \to 4} (f \cdot g)(x) = \left(\lim_{x \to 4} f(x) \right) \cdot \left(\lim_{x \to 4} g(x) \right) = -3 \cdot 13/2.$$

Discussion.   Let $\epsilon > 0$. We want to find $\delta > 0$ such that

$$\text{if} \quad 0 < |x - 4| < \delta, \quad \text{then} \quad \left| (2x - 11) \cdot \frac{2x + 5}{x - 2} - (-3 \cdot 13/2) \right| < \epsilon.$$

We again want to isolate the terms

$$|(2x - 11) + 3| \quad \text{and} \quad \left| \frac{2x + 5}{x - 2} - \frac{13}{2} \right|.$$

Can you explain why we want to do this?

    We can achieve a partial isolation of the terms by modifying the difference. We can do this without changing the value of the difference. Do you recall how this was done for the product of two convergent sequences in Lab 6? If we add and then subtract the product $(2x - 11) \cdot \frac{13}{2}$, we can apply the distributive property to write a sum so that the terms

$$|(2x - 11) + 3| \quad \text{and} \quad \left| \frac{2x + 5}{x - 2} - \frac{13}{2} \right|$$

appear:

$$\left|(2x - 11) \cdot \frac{2x + 5}{x - 2} - \left(-3 \cdot \frac{13}{2}\right)\right|$$

$$= \left|(2x - 11) \cdot \frac{2x + 5}{x - 2} - (2x - 11) \cdot \frac{13}{2} + (2x - 11) \cdot \frac{13}{2} - \left(-3 \cdot \frac{13}{2}\right)\right|$$

$$= \left|(2x - 11)\left(\frac{2x + 5}{x - 2} - \frac{13}{2}\right) + \frac{13}{2}\left[(2x - 11) + 3\right]\right|$$

$$\leq |2x - 11|\left|\frac{2x + 5}{x - 2} - \frac{13}{2}\right| + \frac{13}{2}\left|(2x - 11) + 3\right|. \tag{$*$}$$

The problem in $(*)$ is the presence of the nonconstant term $|2x - 11|$ in the first summand. However, since the function $f(x) = 2x - 11$ has a limit at $x = 4$, we can, as you will see below, place a bound on this term, so that the definition can be applied to the other factor.

Since we want the entire sum to be less than ϵ, we would like to make each summand less than $\epsilon/2$. Because the factor in the second summand is multiplied by $13/2$, we want $\left|(2x - 11) + 3\right|$ to be less than $2\epsilon/26$. Go back to Section 11.2.1 and see what modifications, if any, need to be made to find $\delta_1 > 0$ such that

$$\text{if} \quad 0 < |x - 4| < \delta_1, \quad \text{then} \quad \left|(2x - 11) + 3\right| < \frac{2\epsilon}{26}.$$

For the first summand, we find a bound for $|2x - 11|$. The bound cannot be expressed in terms of an unspecified value of ϵ. Consequently, we select a specific value of ϵ, say $\epsilon = 1$, and find a $\delta_2 > 0$ that ensures that $|2x - 11 - (-3)| < 1$. Rewriting this inequality places a bound on $|2x - 11|$ for all x such that $0 < |x - 4| < \delta_2$. Before continuing, find a suitable δ_2, rewrite the inequality, and then identify the bound, which, for the remainder of this discussion, we will call B.

Now that we have found a bound for $|2x - 11|$, our next task is to deal with the term $\left|\frac{2x+5}{x-2} - \frac{13}{2}\right|$. Since the limit of g exists at $x = 4$, we can find, given $\epsilon/2B$, $\delta_3 > 0$ such that

$$\text{if} \quad 0 < |x - 4| < \delta_3, \quad \text{then} \quad \left|\frac{2x + 5}{x - 2} - \frac{13}{2}\right| < \frac{\epsilon}{2B}.$$

Use the results of Section 11.2.2 to help you find a suitable δ_3. Why do we want the term to be less than $\epsilon/2B$?

Proof. Let $\epsilon > 0$. Let $\delta = ?$ (*you are asked to determine this*). Choose x such that $0 < |x-4| < \delta$. Then,

$$\left|(2x - 11) \cdot \frac{2x + 5}{x - 2} - \left(-3 \cdot \frac{13}{2}\right)\right| < \epsilon.$$

Q.E.D.

Exercises

1. Complete the proof above by giving the exact form for δ and filling in the missing details of the proof.

2. Suppose h and k are two functions such that $\lim_{x \to x_0} h(x) = L$ and $\lim_{x \to x_0} k(x) = M$. Use the $\epsilon - \delta$ definition of limit, together with the techniques displayed in the example above, to prove that $\lim_{x \to x_0} (h \cdot k)(x) = L \cdot M$.

3. If h and k are both continuous at $x = x_0$, what changes would you have to make in the proof you wrote in the previous exercise to prove the continuity of $h \cdot k$ at $x = x_0$? In Lab 9, you used the sequence definition to prove that the product of two continuous functions is also continuous. Write a similar proof using the $\epsilon - \delta$ definition of continuity.

11.3.3   The Quotient of Two Functions

In this case, we work a bit differently than we did in trying to prove theorems involving the sum and product. Rather than work with two specific functions, we consider the reciprocal of a single arbitrary function. Specifically, if h is a function that is nonzero in an interval about x_0 and that has a nonzero limit M at $x = x_0$, we show that

$$\lim_{x \to x_0} \frac{1}{h(x)} = \frac{1}{\lim\limits_{x \to x_0} h(x)} = \frac{1}{M}.$$

Why do you suppose M is required to be nonzero? Why must h be nonzero in an interval about $x = x_0$?

Discussion.   Let $\epsilon > 0$. We want to find $\delta > 0$ such that

$$\text{if} \quad 0 < |x - x_0| < \delta, \quad \text{then} \quad \left| \frac{1}{h(x)} - \frac{1}{M} \right| < \epsilon.$$

In order to get an expression involving $\left| h(x) - M \right|$, one to which we can apply the definition, we first find a common denominator:

$$\left| \frac{1}{h(x)} - \frac{1}{M} \right| = \left| \frac{M - h(x)}{Mh(x)} \right|$$
$$= \frac{|h(x) - M|}{|M||h(x)|}.$$

Although we get $\left| h(x) - M \right|$ in the numerator, we have to deal with the variable term $|h(x)|$ in the denominator. We encountered a similar problem in Section 11.3.2. As with the case of the product, we can construct a bound for the term $1/|h(x)|$. We do not want this bound to involve an unspecified value of ϵ. Consequently, we fix a value of ϵ. In this case, let $\epsilon = |M|/2$. Why do we make this particular choice? Since $\lim_{x \to x_0} h(x) = M$, there exists $\delta_1 > 0$ such that

$$\text{if} \quad 0 < |x - x_0| < \delta_1, \quad \text{then} \quad |h(x) - M| < \frac{|M|}{2}.$$

As the graph shows, the choice of $|M|/2$ for ϵ ensures that $h(x) \neq 0$ for all x within δ_1 units of x_0, with the possible exception of x_0 itself. Why is this important?

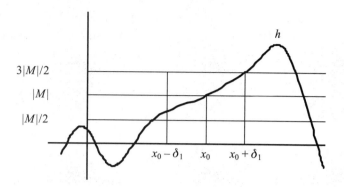

As you study the steps below, identify that step where $h(x) \neq 0$ for all $0 < |x - x_0| < \delta_1$ plays a critical role:

$$|M| = |M - h(x) + h(x)|$$
$$\leq |M - h(x)| + |h(x)|$$
$$= |h(x) - M| + |h(x)|$$
$$< \frac{|M|}{2} + |h(x)| \quad \Longrightarrow$$
$$\frac{|M|}{2} < |h(x)| \quad \Longrightarrow$$
$$\frac{1}{|h(x)|} < \frac{2}{|M|}, \quad \text{where } 0 < |x - x_0| < \delta_1.$$

Now that we have a bound for $1/h(x)$, we can apply the definition to the term $|h(x) - M|$. Since $\lim_{x \to x_0} h(x) = M$, we can find $\delta_2 > 0$ such that

$$\text{if} \quad 0 < |x - x_0| < \delta_2, \quad \text{then} \quad |h(x) - M| < \frac{|M|^2 \epsilon}{2}.$$

Why do we work with such a seemingly complicated expression involving ϵ?

Exercises

1. Use what you learned in the preceding discussion to write a formal proof: If $\lim_{x \to x_0} h(x) = M$, where $M \neq 0$ and h is nonzero in an interval about x_0, then

$$\lim_{x \to x_0} \frac{1}{h(x)} = \frac{1}{M}.$$

2. Suppose that l and k are two functions such that $\lim_{x \to x_0} l(x) = L$ and $\lim_{x \to x_0} k(x) = M$, where $M \neq 0$ and k is nonzero in an interval about x_0. Prove that the limit of the quotient exists. Specifically, show that

$$\lim_{x \to x_0} \frac{l(x)}{k(x)} = \frac{L}{M}.$$

3. What changes would you have to make in your proof in exercise 1 to prove the continuity of $1/h$? Using the $\epsilon - \delta$ definition of continuity, show that if h is continuous at $x = x_0$ and $h(x_0) \neq 0$, then $1/h$ is continuous at $x = x_0$.

11.4   Questions for Reflection

1. Is there a redundancy in the statements given in exercises 1 and 2 of Section 11.3.3? Specifically, is it sufficient to assume that $M \neq 0$? Why, or why not?

2. If $\lim_{x \to x_0} g(x) = M$ for some function g, use the $\epsilon - \delta$ definition of limit to show that there exists $\delta > 0$ such that $|g(x)| < |M| + 1$ for all x such that $0 < |x - x_0| < \delta$.

3. Determine which of the following is true. For the statement that is false, provide a counterexample. For the statement that is true, provide a proof. For each statement, assume that f is a function.

   (a) If $\lim_{x \to x_0} f(x)$ exists, then f is continuous at $x = x_0$.

   (b) If f is continuous at $x = x_0$, then $\lim_{x \to x_0} f(x)$ exists.

4. Use the $\epsilon - \delta$ definitions of limit and continuity to help you prove the following statement: A function f is continuous at a point $x = x_0$ if and only if $\lim_{x \to x_0} f(x) = f(x_0)$. Does this statement look familiar to you? What is the relationship between this biconditional statement (if and only if statement) and the definition discussed in Lab 9, Section 9.2?

5. Use the definition of limit to prove the Squeeze Theorem. Let f, g, and h be three functions such that $f(x) \leq g(x) \leq h(x)$ for all x in an interval about x_0. If

$$\lim_{x \to x_0} f(x) = L = \lim_{x \to x_0} h(x),$$

where L is a real number, show that

$$\lim_{x \to x_0} g(x) = L.$$

State and prove a similar theorem if f, g, and h are assumed to be continuous at x_0.

6. If $\lim_{x \to x_0} g(x) = a$, and if f is continuous at $x = a$, show that

$$\lim_{x \to x_0} f(g(x)) = f(a) = f\left(\lim_{x \to x_0} g(x) \right).$$

7. Define a function g by $g(x) = x$. Use the $\epsilon - \delta$ definition of continuity to show that g is continuous for all $x \in \mathbb{R}$. Use this result to show that the function h defined by $h(x) = x^n$, $n \in \mathbb{N}$, is continuous for all $x \in \mathbb{R}$. How could you use this result, together with other theorems considered in this lab, to prove the continuity of a function defined by a polynomial?

8. Find or construct examples of functions that satisfy the given conditions:

(a) A function h that can be expressed as the composition of two functions, say $h(x) = (g \circ f)(x)$, such that

$$\lim_{x \to x_0} (g \circ f)(x) \neq g \left(\lim_{x \to x_0} f(x) \right).$$

(b) A function r that can be expressed as the composition of two functions, say $r(x) = (t \circ s)(x)$, such that

$$\lim_{x \to x_0} (t \circ s)(x) = t \left(\lim_{x \to x_0} s(x) \right),$$

but such that s is not continuous at $x = x_0$.

12

Uniform Convergence of a Sequence of Functions

12.1 Introduction

Thus far, we have examined sequences with constant terms; that is, the term itself is not a (nonconstant) function. However, it is possible to create a sequence where each term is itself a (nonconstant) function. This construction is aptly called a *sequence of functions*. Taking the limit of such a sequence is fairly straightforward. The limit is a function whose domain is a subset of the domain common to all the terms of the sequence. To find the value of the limit function at a point x_0 in the domain, one proceeds as follows:

- evaluate each function at that point; this produces a sequence of constant terms $(f_n(x_0))_{n=1}^{\infty}$;
- take the limit of this sequence.

This limit value, if finite, is the value of the limit function at that point x_0. We call the limit function obtained in this way *the pointwise limit*. We demonstrate this process in the examples in this lab.

There are two types of convergence related to sequences of functions. In this lab, we explore both of these types of convergence. It should be obvious to you when you look at the computer output that one type of convergence is stronger than the other. This stronger type of convergence is called *uniform convergence*.

Natural questions arise when one looks at sequences of functions. Examples of questions include:

- If each term of a sequence of functions is continuous, is the limit function also continuous?
- If one computes the limit of a sequence of integrals of functions, is the result equal to the integral of the limit function? That is, does the following equality hold:

$$\lim_{n \to \infty} \int_a^b f_n(x)\, dx = \int_a^b \lim_{n \to \infty} f_n(x)\, dx?$$

- If each term of a sequence is a differentiable function, is the limit function differentiable?

These are just three examples of questions. The answer to these questions hinges upon the type of convergence. In this lab, we explore the first of these questions.

For this lab, you may wish to use the accompanying *Visual Guide for Lab 12*, which can be found at www.saintmarys.edu/~jsnow. The *Visual Guide* also appears in the appendix.

12.2 Using Examples to Understand Pointwise Limits

1. Consider the following sequence of functions: $(f_n(x) = x^n)_{n=1}^{\infty}$ on the interval $[0, 1]$.

    (a) Graph the functions: $f_1(x) = x$, $f_3(x) = x^3$, $f_5(x) = x^5$, $f_{10}(x) = x^{10}$, $f_{15}(x) = x^{15}$ on the same graph with domain $[0, 1]$.

    If you fix a particular point x_0 in the interval $[0, 1]$, then you can generate a sequence of real numbers $(f_n(x_0))_{n=1}^{\infty}$. The terms of the sequence are just the values generated by each function f_n when it is evaluated at x_0. You can think of this sequence as the set of y-coordinates of the points on the curves with x-coordinate $x = x_0$. We demonstrate this process algebraically and graphically in the questions below.

    (b) Let $x_0 = .5$. Compute the first 6 terms of the sequence $(f_n(.5))_{n=1}^{\infty}$. What does $\lim_{n \to \infty} f_n(.5)$ appear to be?

    (c) Now look at the graph you created in part (a). Does the graph support your answer to part (b)? Remember the terms of this sequence $(f_n(.5))_{n=1}^{\infty}$ are represented in the graph as points of the form $(.5, f_n(.5))_{n=1}^{\infty}$.

    (d) Let $x_0 = .2$. Compute the first 6 terms of the sequence $(f_n(.2))_{n=1}^{\infty}$. What does $\lim_{n \to \infty} f_n(.2)$ appear to be?

    (e) Using the graph, what does $\lim_{n \to \infty} f_n(.2)$ appear to be?

    (f) Let $x_0 = .7$. Compute the first 6 terms of the sequence $(f_n(.7))_{n=1}^{\infty}$. What does $\lim_{n \to \infty} f_n(.7)$ appear to be?

    (g) Using the graph, what does $\lim_{n \to \infty} f_n(.7)$ appear to be?

    (h) Using the graph, for any $x_0 \neq 1$, what does $\lim_{n \to \infty} f_n(x_0)$ appear to be?

    (i) Using the graph, what does $\lim_{n \to \infty} f_n(1)$ appear to be? Since $(f_n(x))_{n=1}^{\infty}$ converges for any $x \in [0, 1]$, we can define a new function which we will denote by f and which we will call the *pointwise limit of* $(f_n)_{n=1}^{\infty}$ on $[0, 1]$. We define

    $$f(x) = \lim_{n \to \infty} f_n(x).$$

    (j) What is f in this case?

    (k) Graph the function f.

    (l) What can you say about the continuity of each f_n on the interval $[0, 1]$?

    (m) What can you say about the continuity of f on the interval $[0, 1]$?

2. We repeat the above exercises with another sequence of functions:

    $$\left(f_n(x) = \frac{x^n}{1 + x^n} \right)_{n=1}^{\infty}$$

    on the interval $[0, 2]$.

(a) Graph the functions:

$$f_1(x) = \frac{x}{1+x}, \quad f_{10}(x) = \frac{x^{10}}{1+x^{10}}, \quad f_{50}(x) = \frac{x^{50}}{1+x^{50}},$$

$$f_{100}(x) = \frac{x^{100}}{1+x^{100}}, \quad f_{300}(x) = \frac{x^{300}}{1+x^{300}}$$

on the same graph with domain $[0, 2]$. Again, if you fix a particular point x_0 in the interval $[0, 2]$, then you can generate a sequence of real numbers $(f_n(x_0))_{n=1}^{\infty}$. The terms of the sequence are just the values generated by each function f_n when it is evaluated at x_0. Recall that you can think of this sequence as the set of y-coordinates of the points on the curves with x-coordinate $x = x_0$.

(b) Let $x_0 = .5$. Compute the first 6 terms of the sequence $(f_n(.5))_{n=1}^{\infty}$. What does $\lim_{n\to\infty} f_n(.5)$ appear to be?

(c) Now look at the graph you created in part (a). Does the graph support your answer to part (b)? Remember the terms of this sequence $(f_n(.5))_{n=1}^{\infty}$ are represented on the graph as points of the form $(.5, f_n(.5))$.

(d) Let $x_0 = .4$. Compute the first 6 terms of the sequence $(f_n(.4))_{n=1}^{\infty}$. What does $\lim_{n\to\infty} f_n(.4)$ appear to be?

(e) Using the graph, what does $\lim_{n\to\infty} f_n(.4)$ appear to be?

(f) Let $x_0 = .1$. Compute the first 6 terms of the sequence $(f_n(.1))_{n=1}^{\infty}$. What does $\lim_{n\to\infty} f_n(.1)$ appear to be?

(g) Using the graph, what does $\lim_{n\to\infty} f_n(.1)$ appear to be?

(h) Using the graph, for any $x_0 \in [0, 1)$, what does $\lim_{n\to\infty} f_n(x_0)$ appear to be?

(i) Let $x_0 = 1$. Compute the first 6 terms of the sequence $(f_n(1))_{n=1}^{\infty}$. What does $\lim_{n\to\infty} f_n(1)$ appear to be?

(j) Using the graph, what does $\lim_{n\to\infty} f_n(1)$ appear to be?

(k) Let $x_0 = 1.2$. Compute the first 6 terms of the sequence $(f_n(1.2))_{n=1}^{\infty}$. What does $\lim_{n\to\infty} f_n(1.2)$ appear to be?

(l) Using the graph, what does $\lim_{n\to\infty} f_n(1.2)$ appear to be?

(m) Using the graph, what does $\lim_{n\to\infty} f_n(1.4)$ appear to be?

(n) Using the graph, what does $\lim_{n\to\infty} f_n(1.8)$ appear to be?

(o) Using the graph, what does $\lim_{n\to\infty} f_n(2)$ appear to be?

(p) Using the graph, for any $x_0 \in (1, 2]$, what does $\lim_{n\to\infty} f_n(x_0)$ appear to be?

(q) As before, we can compute the pointwise limit f of $(f_n)_{n=1}^{\infty}$ on $[0, 2]$. What is f in this case?

(r) Graph the function f.

(s) What can you say about the continuity of each f_n on the interval $[0, 2]$?

(t) What can you say about the continuity of f on the interval $[0, 2]$?

3. We repeat this process yet again with another sequence of functions:

$$\left(f_n(x) = \frac{x}{1+nx^2}\right)_{n=1}^{\infty}$$

on the interval $[0, 1]$.

(a) Graph the functions:

$$f_1(x) = \frac{x}{1+x^2}, \quad f_{10}(x) = \frac{x}{1+10x^2}, \quad f_{50}(x) = \frac{x}{1+50x^2},$$

$$f_{100}(x) = \frac{x}{1+100x^2}, \quad f_{300}(x) = \frac{x}{1+300x^2}$$

on the same graph.

(b) Let $x_0 = .5$. Compute the first 20 terms of the sequence $(f_n(.5))_{n=1}^{\infty}$. What does $\lim_{n\to\infty} f_n(.5)$ appear to be?

(c) Does the graph support your answer to part (b)?

(d) Using the graph, what does $\lim_{n\to\infty} f_n(.2)$ appear to be?

(e) Using the graph, what does $\lim_{n\to\infty} f_n(1)$ appear to be?

(f) Using the graph, for any $x_0 \in [0, 1]$, what does $\lim_{n\to\infty} f_n(x_0)$ appear to be?

(g) What is the pointwise limit f in this case?

(h) Graph the function f.

(i) What can you say about the continuity of each f_n on the interval $[0, 1]$?

(j) What can you say about the continuity of f on the interval $[0, 1]$?

12.3   Understanding the Two Types of Convergence

There is a difference between the convergence of the sequences of functions in questions 1 and 2 and the sequence in question 3. We try to formulate that distinction now.

1. To the graph you created in part (a) of question 1 in Section 12.2, we add three functions: the pointwise limit function f, the function $f + \epsilon$, and the function $f - \epsilon$, where we let ϵ take on different positive values.

(a) Let $\epsilon = .5$. Graph the functions f, $f + \epsilon$, and $f - \epsilon$.

   i. Does it appear that the particular sequence function f_{10} fits **completely** in the bands created by $f + \epsilon$ and $f - \epsilon$? That is, for **all** $x \in [0, 1]$, is it true that $f_{10}(x) \in (f(x) - \epsilon, f(x) + \epsilon)$? If for some $x \in [0, 1]$, you have $f_{10}(x) \notin (f(x) - \epsilon, f(x) + \epsilon)$, identify those points.

   ii. Does it appear that the particular sequence function f_{15} fits **completely** in the bands created by $f + \epsilon$ and $f - \epsilon$? That is, for **all** $x \in [0, 1]$, is it true that $f_{15}(x) \in (f(x) - \epsilon, f(x) + \epsilon)$? If for some $x \in [0, 1]$, you have $f_{15}(x) \notin (f(x) - \epsilon, f(x) + \epsilon)$, identify those points.

   iii. If your answer to (i) or (ii) was no, do you think the answer would be yes if one considered f_n, where n is taken sufficiently large? Experiment with different f_n.

(b) Let $\epsilon = .2$. Repeat part (a) i, ii, iii for this value of ϵ.

(c) Let $\epsilon = .1$. Repeat part (a) i, ii, iii for this value of ϵ.

2. To the graph you created in part (a) of question 2 in Section 12.2, we add three functions: the pointwise limit function f, the function $f + \epsilon$, and the function $f - \epsilon$, where we let ϵ take on different positive values. To make our point easier for you to see, adjust the domain interval to the subinterval $[.95, 1.05]$.

(a) Let $\epsilon = .3$. Graph the functions f, $f + \epsilon$, and $f - \epsilon$.

   i. Does it appear that the particular sequence function f_{100} fits **completely** in the bands created by $f + \epsilon$ and $f - \epsilon$? That is, for **all** $x \in [.95, 1.05]$, is it true that $f_{100}(x) \in (f(x)-\epsilon, f(x)+\epsilon)$? If for some $x \in [.95, 1.05]$, you have $f_{100}(x) \notin (f(x)-\epsilon, f(x)+\epsilon)$, identify those points.

   ii. Does it appear that the particular sequence function f_{300} fits **completely** in the bands created by $f + \epsilon$ and $f - \epsilon$? That is, for **all** $x \in [.95, 1.05]$, is it true that $f_{300}(x) \in (f(x)-\epsilon, f(x)+\epsilon)$? If for some $x \in [.95, 1.05]$, you have $f_{300}(x) \notin (f(x)-\epsilon, f(x)+\epsilon)$, name such an x.

   iii. If your answer to (i) or (ii) was no, do you think the answer would be yes if one considered f_n, where n is taken sufficiently large? Experiment with different f_n.

(b) Let $\epsilon = .2$. Repeat part (a) i, ii, iii for this value of ϵ.

(c) Let $\epsilon = .1$. Repeat part (a) i, ii, iii for this value of ϵ.

3. To the graph you created in part (a) of question 3 in Section 12.2, we add three functions: the pointwise limit function f, the function $f + \epsilon$, and the function $f - \epsilon$, where we let ϵ take on different positive values.

(a) Let $\epsilon = .3$. Graph the functions f, $f + \epsilon$, and $f - \epsilon$.

   i. Does it appear that the particular sequence function f_{100} fits **completely** in the bands created by $f + \epsilon$ and $f - \epsilon$? That is, for **all** $x \in [0, 1]$, is it true that $f_{100}(x) \in (f(x) - \epsilon, f(x) + \epsilon)$? If for some $x \in [0, 1]$, you have $f_{100}(x) \notin (f(x) - \epsilon, f(x) + \epsilon)$, identify those points.

   ii. Does it appear that the particular sequence function f_{300} fits **completely** in the bands created by $f + \epsilon$ and $f - \epsilon$? That is, for **all** $x \in [0, 1]$, is it true that $f_{300}(x) \in (f(x) - \epsilon, f(x) + \epsilon)$? If for some $x \in [0, 1]$, you have $f_{300}(x) \notin (f(x) - \epsilon, f(x) + \epsilon)$, identify those points.

   iii. If your answer to (i) or (ii) was no, do you think the answer would be yes if one considered f_n, where n is taken sufficiently large? Experiment with different f_n.

(b) Let $\epsilon = .1$. Repeat part (a) i, ii, iii for this value of ϵ.

(c) Let $\epsilon = .05$. Repeat part (a) i, ii, iii for this value of ϵ.

12.4 Critical Thinking Questions

1. Examining the results of the last section, explain in your own words the difference between the behavior of the sequences from questions 1 and 2 versus the sequence from question 3. The kind of convergence that we see in the sequence from question 3 is special. It is called *uniform convergence*. The idea is that for a given ϵ, there is a common N such that

$$\text{if } \quad n > N, \quad \text{then} \quad |f_n(x) - f(x)| < \epsilon \quad \text{for all } x \text{ in the domain of } f.$$

2. In your graph for question 3 with $\epsilon = .1$, can you tell what N produces $|f_n(x) - f(x)| < \epsilon$ for all x in the domain?

3. Let's look at the graph of another sequence of functions which exhibits this special kind of convergence. Consider the following sequence of functions: $(f_n(x) = nxe^{-n^2x})_{n=1}^{\infty}$ on the interval $[0, 1]$.

   (a) Graph the functions: $f_1(x)$, $f_2(x)$, $f_3(x)$, $f_5(x)$, $f_6(x)$, $f_{10}(x)$, $f_{11}(x)$ on the same graph.

   (b) Find the pointwise limit. Call it f.

   (c) Now add the graphs of f, $f + .1$, $f - .1$ to the graphs of the sequence.

   (d) For what N does it appear that $|f_n(x) - f(x)| < .1$ for all x in the specified domain and for all $n > N$?

   (e) What can you say about the continuity of each f_n on the interval $[0, 1]$?

   (f) What can you say about the continuity of f on the interval $[0, 1]$?

12.5 Questions for Reflection

1. Explain in your own words why we say uniform convergence is stronger than simple pointwise convergence.

2. Now we want to see the power of uniform convergence. Complete the table below, answering yes or no to each question at the head of each column.

Sequence	Is f_n continuous for all n?	Is the pointwise limit continuous?	Is the convergence uniform?
$(f_n(x) = x^n)_{n=1}^{\infty}$			
$\left(f_n(x) = \dfrac{x^n}{1 + x^n} \right)_{n=1}^{\infty}$			
$\left(f_n(x) = \dfrac{x}{1 + nx^2} \right)_{n=1}^{\infty}$			
$\left(f_n(x) = nxe^{-nx^2} \right)_{n=1}^{\infty}$			

From the results in the table, does it appear that uniform convergence is sufficient to guarantee that the limit function is continuous?

3. Does your answer to the previous question give you an easy way to conclude a condition by which a sequence of functions does not converge uniformly?

4. Suppose that the sequence of functions $(f_n)_{n=1}^{\infty}$ converges uniformly to f on a set S. Let $c \in \mathbb{R}$. Show that the sequence $(cf_n)_{n=1}^{\infty}$ converges uniformly to cf on S.

5. Suppose that a sequence of functions $(f_n)_{n=1}^{\infty}$ converges uniformly to f and that a sequence of functions $(g_n)_{n=1}^{\infty}$ converges uniformly to g on a set S. Show that the sequence $(f_n + g_n)_{n=1}^{\infty}$ converges uniformly to $f + g$ on S.

Appendix: Visual Guides

For those labs where the use of technology proved to be beneficial, we have written *Visual Guides* using *Maple* code. The *Visual Guides* provide suggested sequences of *Maple* commands for those exercises requiring the analysis of graphs. With suitable modifications, the code can be used to graph each example. In addition to appearing in the appendix, each *Visual Guide* can also be found online at www.saintmarys.edu/~jsnow. Periodic updates and refinements can be found by checking the website.

As an alternative to the *Visual Guides, Maplets* can also be found at www.saintmarys.edu/~jsnow.

Visual Guide for Lab 3

Our goal in this exploration is to use the power of graphics to motivate the definition of the limit of a sequence. A sequence is denoted by $(s(n))$. The nth term of the sequence is denoted $s(n)$. Points on the graph of the sequence are denoted by the usual ordered pair notation $(n, s(n))$. You should think of the code in this worksheet as a template. You should feel free to modify the code as needed. A section of code is devoted to Example Set 1. Another section is devoted to Example Set 2.

Load the Plotting Package

You need to load the following package in order to plot.

```
> with(plots):
Warning, the name arrow has been redefined
```

Section 2, Example Set 1

The code in this section is for Example Set 1 of Section 2. Before you begin the first exercise of the lab, you should look at the first five parts of this section of the worksheet.

Define the Sequences of Example Set 1

In the first section of the lab, you are considering two sequences: $(a(n)) = ((-1)^n/n)$ and $(b(n)) = (((-1)^n)*(2+(1/n)))$. So in this section of the code, we first define the sequences that we are investigating. Note that we have used a decimal representation of one of the real numbers in the formula for the sequence. In this format, evaluation of the terms of the sequence will be as decimals.

```
> a:=n->(-1.)^n/n;
> b:=n->((-1)^n)*(2.+1/n);
```

$$a := n \mapsto \frac{(-1.0)^n}{n}$$

$$b := n \mapsto (-1)^n \left(2.0 + \frac{1}{n}\right)$$

Enter the Indexing Set for Each Sequence

Before you graph a sequence, you need to determine that part of the domain and range that will be displayed. You do this by first generating some numerical data on the sequence. If $s(n)$ represents the general term of the sequence, you might first want to look at the points $(1, s(1)), (2, s(2),)(3, s(3)), \ldots, (20, s(20))$. Those points you consider will be determined by your need. For example, to estimate long-range behavior of the sequence, you may discover that you need to see more points of the sequence than are currently shown. You might want to zoom in on the behavior of the sequence between $n = 15$ and $n = 50$. Based on your need, you adjust the parameters below. Remember that the indexing set is just a subset of the domain of the sequence.

For the sequence $(a(n))$, set the maximum value for the indexing set. We have initially set this max at 20. We count by ones (the step size) and start the indexing set at 1. You could count by 2's or any other positive integer value. The numbers—nmaxa, the starting value, and the step size—may be changed, if you find the need or have the desire.

```
> nmaxa:=20:
> Xa:=1:
> for i from 2 to nmaxa by 1 do
> Xa:=Xa,i;
> od:
```

For the sequence $(b(n))$, set the maximum value for the indexing set. We have initially set this max at 20. We count by ones (the step size) and start the indexing set at 1. You could count by 2's or any other positive integer value. The numbers—nmaxb, the starting value, and the step size—may be changed, if you find the need or have the desire.

```
> nmaxb:=20:
> Xb:=1:
> for i from 2 to nmaxb by 1 do
> Xb:=Xb,i;
> od:
```

Generating Terms of the Sequences

We generate some of the points of the sequence for the indexing set defined above. This will help us later in setting the limits on the y-range of the graph.

```
> A:=seq([n,a(n)],n=Xa);
```

$$A := [1, -1.0], [2, 0.5000000000], [3, -0.3333333333], [4, 0.2500000000], [5, -0.2000000000],$$
$$[6, 0.1666666667], [7, -0.1428571429], [8, 0.1250000000], [9, -0.1111111111],$$
$$[10, 0.1000000000], [11, -0.09090909091], [12, 0.08333333333], [13, -0.07692307692],$$
$$[14, 0.07142857143], [15, -0.06666666667], [16, 0.06250000000], [17, -0.05882352941],$$
$$[18, 0.05555555556], [19, -0.05263157895], [20, 0.05000000000]$$

```
> B:=seq([n,b(n)],n=Xb);
```

$B := [1, -3.0], [2, 2.500000000], [3, -2.333333333], [4, 2.250000000], [5, -2.200000000],$

$\qquad [6, 2.166666667], [7, -2.142857143], [8, 2.125000000], [9, -2.111111111],$

$\qquad [10, 2.100000000], [11, -2.090909091], [12, 2.083333333], [13, -2.076923077],$

$\qquad [14, 2.071428571], [15, -2.066666667], [16, 2.062500000], [17, -2.058823529],$

$\qquad [18, 2.055555556], [19, -2.052631579], [20, 2.050000000]$

Set the Range on the y-values for Graphing

From the numerical calculations above, we can determine a range on the y-values for the graphs of the sequences. We do this first for sequence $(a(n))$ and then for sequence $(b(n))$.

```
> arange:=-1..1;
> brange:=-3..2.5;
```

$$arange := -1 \ldots 1$$

$$brange := -3 \ldots 2.5$$

Plot the Two Sequences of Example Set 1

Now we plot the first sequence for the indexing set we have defined. We set the n-range at 0 to nmaxa. For the y-range, we use the settings above. Then we store the graph of the first sequence in a plot structure called "gran."

```
> plot([A],n=0..nmaxa,y=arange,style=point,symbol=DIAMOND);
```

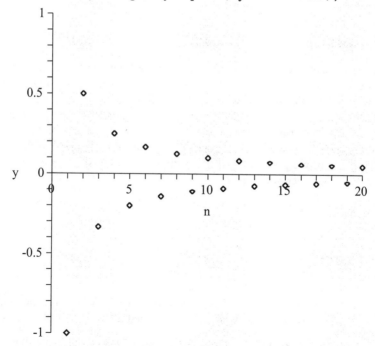

```
> gran:=plot([A],n=0..nmaxa,y=arange,style=point,symbol=DIAMOND):
```

Now we plot the second sequence for the indexing set we have defined. We set the n-range at 0 to nmaxb. For the y-range, we use the settings above. Then we store the graph of that sequence in a plot structure called "grbn."

```
> plot([B],n=0..nmaxb,y=brange,style=point,symbol=DIAMOND);
```

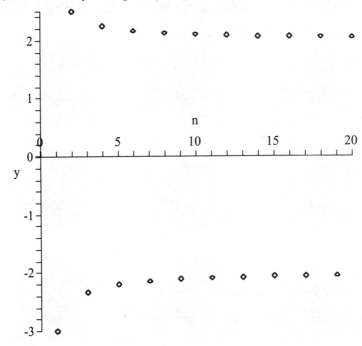

```
> grbn:=plot([B],n=0..nmaxb,y=brange,style=point,symbol=DIAMOND):
```

The epsilon-N Exploration for Example Set 1

Now we are ready to explore graphically the definition of the limit of a sequence. Remember that we say that L is the limit of the sequence $(s(n))$ if the terms of the sequence are getting closer and closer to L. More precisely, we first issue a challenge: we specify how close to L we want the terms of the sequence to be. This measure of closeness we call epsilon. The limit of Sequence 1 is denoted La and that of Sequence 2 is denoted Lb. We know the terms of the sequence are epsilon-close to L if on the graph of the sequence, the points fall within the lines $y = L - $ epsilon and $y = L + $ epsilon. Let us call those lines the boundary lines and name the graph of the pair for Sequence 1 "linea" and for Sequence 2 "lineb." We first do the graphing for Sequence 1 and then for Sequence 2.

For Sequence 1:
We set a value for epsilon.

```
> epsilon:=.5:
```

We fill in the value for the limit.

```
> La:=0:
```

We store in "linea" the graph of the boundary lines for Sequence 1. We display the graph of the sequence and the boundary lines. Do you see that we have easily trapped almost all of the points?

```
> linea:=plot({La-epsilon,La+epsilon},n=0..nmaxa,color= blue):
> display(gran,linea);
```

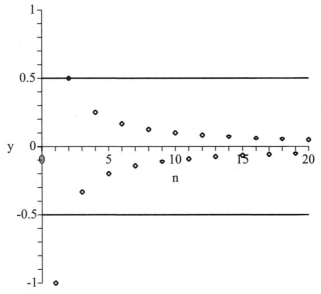

For Sequence 2:
We set a value for epsilon.

```
> epsilon:=.5:
```

We fill in the value for L.

```
> Lb:=2:
```

We store in "lineb" the graph of the boundary lines for Sequence 2. We display the graph of the sequence and the boundary lines.

```
> lineb:=plot({Lb-epsilon,Lb+epsilon},n=0..nmaxb,color = blue):
> display(grbn,lineb);
```

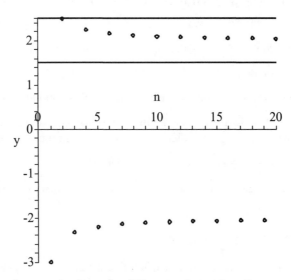

Now you will adjust the above code above for different values of epsilon.

Section 2, Set 2

The code in this section is for Example Set 2 of Section 2. We enter only one sequence. You can adjust the code for the other sequences.

Enter the Indexing Set

For your convenience, we include here a chance to re-define the indexing set for the sequences in Set 2. Set the maximum value for the indexing set. We have initially set this max at 20. We count by ones (the step size) and start the indexing set at 1. You could count by 2's or any other positive integer value. The numbers—nmaxc, the starting value, and the step size—may be changed, if you find the need or have the desire.

```
> nmaxc:=20:
> Y:=1:
> for j from 2 to nmaxc by 1 do
> Y:=Y,j;
> od:
```

Define a Sequence of Example Set 2

We enter the code for Sequence 3: $(n/(n+1))$.

```
> c:=n->(n/(n+1.));
```

$$c := n \mapsto \frac{n}{n+1.0}$$

```
> C:=seq([n,c(n)],n=Y);
```

$$C := [1, 0.5000000000], [2, 0.6666666667], [3, 0.7500000000], [4, 0.8000000000],$$
$$[5, 0.8333333333], [6, 0.8571428571], [7, 0.8750000000], [8, 0.8888888889],$$
$$[9, 0.9000000000], [10, 0.9090909091], [11, 0.9166666667], [12, 0.9230769231],$$
$$[13, 0.9285714286], [14, 0.9333333333], [15, 0.9375000000], [16, 0.9411764706],$$
$$[17, 0.9444444444], [18, 0.9473684211], [19, 0.9500000000], [20, 0.9523809524]$$

We explain below how to handle a piecewise-defined sequence, should the need arise. The code is given for the sequence $(d(n))$ which assigns the value n^2 if n is even and the value 0 if n is odd.

```
> d:=n->piecewise(type(n,even),n^2,type(n,odd),0);
```

$$d := n \mapsto \text{piecewise}(\text{type}(n, even), n^2, \text{type}(n, odd), 0)$$

Set the Range on the y-values for Graphing

From the numerical calculations above, we can determine a range on the y-values for the graph of the sequence. These values can be adjusted as necessary.

```
> crange:=0..1.5;
```

$$crange := 0..1.5$$

Plot the Sequence

Now we plot the sequence for the indexing set we have defined. We set the n-range at 0 to nmaxc. Then we store the graph of the sequence in a plot structure called "grcn."

```
> plot([C],n=0..nmaxc,y=crange,style=point,symbol=DIAMOND);
```

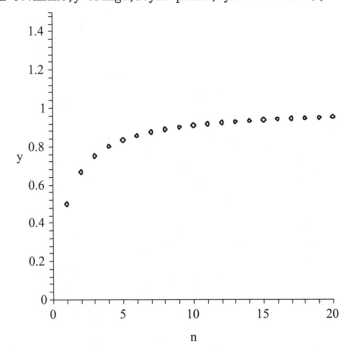

```
> grcn:=plot([C],n=0..nmaxc,y=crange,style=point,symbol=DIAMOND):
```

The epsilon-N exploration for Example Set 2

In this section we again explore graphically the definition of the limit of a sequence. We use the letter L to denote the value we are considering as the limit of the sequence $(s(n))$. We know the terms of the sequence are epsilon-close to L if on the graph of the sequence, the points fall within the lines $y = L - $ epsilon and $y = L + $ epsilon. Again let us call those lines the boundary lines and give the graph of the pair of them the name "linec."

We set a value for epsilon.

```
> epsilon:=.5:
```

We fill in the value we think is the limit.

```
> L:=1:
```

We store in "linec" the graph of the boundary lines for the sequence. We display the graph of the sequence and the boundary lines.

```
> linec:=plot({L-epsilon, L+epsilon},n=0..nmaxc,color=blue):
> display(grcn,linec);
```

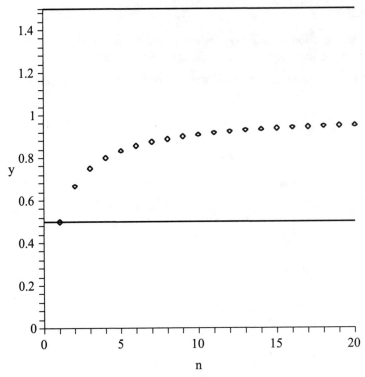

Now you repeat the process for Sequences 4, 5, and 6.

```
> restart;
```

Visual Guide for Labs 4 and 5

Our goal in this exploration is to use the power of graphics to understand the definition of the limit of a sequence. We will examine the meaning of each phrase of this complex definition. You should think of the code in this worksheet as a template. You should feel free to modify the code as each individual sequence suggests.

Load the Plotting Package

Load the plotting package.

```
> with(plots):
Warning, the name arrow has been redefined
```

Before graphing a sequence there are certain tasks to complete: the sequence needs to be defined and the range on the n (domain) and y (range) variables need to be determined. The purpose of the next four sections of code is to accomplish those tasks.

Define the Sequence

We define the first sequence that we are investigating. We consider the sequence $(a(n)) = (1/n)$.

```
> a:=n->1./n;
```

$$a := n \mapsto 1.0\tfrac{1}{n}$$

We explain below how to handle a piecewise-defined sequence, should the need arise. The code is given for the sequence $(b(n))$ which assigns the value n^2 if n is even and the value 0 if n is odd.

```
> b:=n->piecewise(type(n,even),n^2,type(n,odd),0);
```

$$b := n \rightarrow \text{piecewise}(\text{type}(n, even), n^2, \text{type}(n, odd), 0)$$

Enter the Indexing Set

Remember that the indexing set is just that part of the domain of the sequence on which we examine the behavior of the sequence. Set the maximum value for the indexing set. We have initially set this max at

100. We count by twos and start the indexing set at 1. These numbers may be changed, if you find the need or have the desire.

```
> nmax:=100:
> X:=1:
> for i from 3 to nmax by 2 do
> X:=X,i;
> od:
```

Generating Terms of the Sequence

We generate some of the points of the sequence for the indexing set defined above. This will help us later in setting the limits on the y-range of the graph.

```
> A:=seq([n,a(n)],n=X);
```

$A := [1, 1.0], [3, 0.3333333333], [5, 0.2000000000], [7, 0.1428571429], [9, 0.1111111111],$

$[11, 0.09090909091], [13, 0.07692307692], [15, 0.06666666667], [17, 0.05882352941],$

$[19, 0.05263157895], [21, 0.04761904762], [23, 0.04347826087], [25, 0.04000000000],$

$[27, 0.03703703704], [29, 0.03448275862], [31, 0.03225806452], [33, 0.03030303030],$

$[35, 0.02857142857], [37, 0.02702702703], [39, 0.02564102564], [41, 0.02439024390],$

$[43, 0.02325581395], [45, 0.02222222222], [47, 0.02127659574], [49, 0.02040816326],$

$[51, 0.01960784314], [53, 0.01886792453], [55, 0.01818181818], [57, 0.01754385965],$

$[59, 0.01694915254], [61, 0.01639344262], [63, 0.01587301587], [65, 0.01538461538],$

$[67, 0.01492537313], [69, 0.01449275362], [71, 0.01408450704], [73, 0.01369863014],$

$[75, 0.01333333333], [77, 0.01298701299], [79, 0.01265822785], [81, 0.01234567901],$

$[83, 0.01204819277], [85, 0.01176470588], [87, 0.01149425287], [89, 0.01123595506],$

$[91, 0.01098901099], [93, 0.01075268817], [95, 0.01052631579], [97, 0.01030927835],$

$[99, 0.01010101010]$

Set the Range on the y-values for Graphing

From the numerical calculations above, we can determine a range on the y-values for the graph of the sequence. In this case we initially set the upper y-value at 1 and the lower y-value at -1. These values can be adjusted as necessary.

```
> arange:=-1..1;
```

$$arange := -1..1$$

Plot the Sequence

Now we plot the sequence for the indexing set we have defined. We set the n-range at 0 to nmax. Using the graph, one can make a conjecture as to the limit of our sequence. We store the graph of the sequence in a plot structure called "gran."

```
> plot([A],n=0..nmax,y=arange,style=point,symbol=DIAMOND);
```

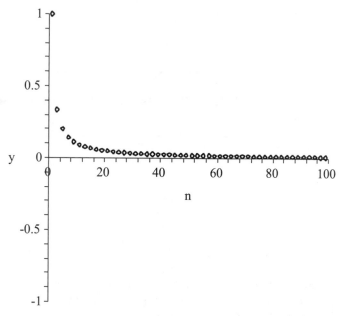

```
> gran:=plot([A],n=0..nmax,y=arange,style=point,symbol=DIAMOND):
```

The epsilon-N exploration

In this section we explore graphically the definition of the limit of a sequence. Remember that we say that L is the limit of the sequence $(a(n))$ if the terms of the sequence are getting closer and closer to L. We first specify epsilon: how close to L we want the terms of the sequence to be. We know the terms of the sequence are epsilon-close to L, if on the graph of the sequence, the points fall within the lines $y = L -$ epsilon and $y = L +$ epsilon. Let us call those lines the boundary lines and give the graph of the pair of them the name "linea." For L to be the limit of the seqeunce, there must be some value, call it N, such that ALL points whose n-coordinate is greater than N will fall within the boundary lines.

We set a value for epsilon. Initially it is .2.

```
> epsilon:=.2:
```

We fill in the value for the limit, based on our graph in the previous part. Next we want to show that the terms of the sequence get epsilon close to L.

```
> L:=0:
```

We define the boundary lines as a plot structure, to be graphed later. We display the graph of the sequence and the boundary lines. Do you see that we have easily trapped almost all of the points?

```
> linea:=plot({L-epsilon, L+epsilon},n=0..nmax,color=blue):
> display(gran,linea);
```

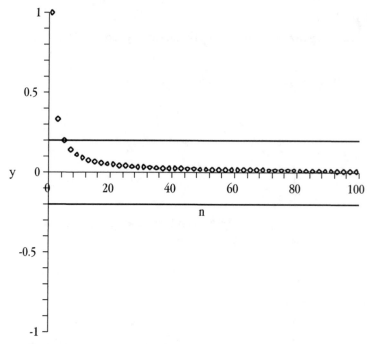

If you need to zoom-in on the *y*-range, go back to the point where you defined arange and replace it with smaller values. Then re-execute the code from that point. If none of the points of the graph is trapped by the boundary lines, you may need to increase nmax.

```
> restart;
```

Visual Guide for Lab 6

Our goal in this exploration is to use the power of graphics to understand algebraic combinations of sequences. You should think of the code in this worksheet as a template. You should feel free to modify the code as each individual example suggests.

Load the Plotting Package

```
> with(plots):
```

Define the Sequences

In the first step of this section of the code we define the sequences that we are investigating. We enter the code for Set 1. The first sequence is $(a(n)) = (1/n)$. The second sequence is $(b(n)) = ((5n+2)/(n+4))$. We let $(c(n))$ be the sum of the sequences $(a(n))$ and $(b(n))$.

```
> a:=n->1./n;
```

$$a := n \mapsto 1.0\frac{1}{n}$$

```
> b:=n->(5.*n-2)/(n+4);
```

$$b := n \mapsto \frac{5.0\,n - 2}{n + 4}$$

```
> c:=n->(a+b)(n);
```

$$c := a + b$$

We explain below how to handle a piecewise-defined sequence, should the need arise. The code is given for the sequence $(d(n))$ which assigns the value n^2 if n is even and the value 0 if n is odd.

```
> d:=n->piecewise(type(n,even),n^2,type(n,odd),0);
```

$$d := n \mapsto \text{piecewise}(\text{type}(n, even), n^2, \text{type}(n, odd), 0)$$

Enter the Indexing Set

Remember that the indexing set is just a subset of the domain of the sequence. Set the maximum value for the indexing set. We have initially set this max at 100. We count by twos and start the indexing set at 1. These numbers may be changed, if you find the need or have the desire.

```
> nmax:=100:
> X:=1:
> for i from 3 to nmax by 2 do
> X:=X,i;
> od:
```

Generating Terms of the Sequences

We generate some of the points of the sequence for the indexing set defined above. This will help us later in setting the limits on the y-range of the graph.

```
> A:=seq([n,a(n)],n=X);
```

$A := [1, 1.0], [3, 0.3333333333], [5, 0.2000000000], [7, 0.1428571429], [9, 0.1111111111],$

$[11, 0.09090909091], [13, 0.07692307692], [15, 0.06666666667], [17, 0.05882352941],$

$[19, 0.05263157895], [21, 0.04761904762], [23, 0.04347826087], [25, 0.04000000000],$

$[27, 0.03703703704], [29, 0.03448275862], [31, 0.03225806452], [33, 0.03030303030],$

$[35, 0.02857142857], [37, 0.02702702703], [39, 0.02564102564], [41, 0.02439024390],$

$[43, 0.02325581395], [45, 0.02222222222], [47, 0.02127659574], [49, 0.02040816326],$

$[51, 0.01960784314], [53, 0.01886792453], [55, 0.01818181818], [57, 0.01754385965],$

$[59, 0.01694915254], [61, 0.01639344262], [63, 0.01587301587], [65, 0.01538461538],$

$[67, 0.01492537313], [69, 0.01449275362], [71, 0.01408450704], [73, 0.01369863014],$

$[75, 0.01333333333], [77, 0.01298701299], [79, 0.01265822785], [81, 0.01234567901],$

$[83, 0.01204819277], [85, 0.01176470588], [87, 0.01149425287], [89, 0.01123595506],$

$[91, 0.01098901099], [93, 0.01075268817], [95, 0.01052631579], [97, 0.01030927835],$

$[99, 0.01010101010]$

```
> B:=seq([n,b(n)],n=X);
```

$B := [1, 0.6000000000], [3, 1.857142857], [5, 2.555555556], [7, 3.0], [9, 3.307692308],$

$[11, 3.533333333], [13, 3.705882353], [15, 3.842105263], [17, 3.952380952],$

$[19, 4.043478261], [21, 4.120000000], [23, 4.185185185], [25, 4.241379310],$

$[27, 4.290322581], [29, 4.333333333], [31, 4.371428571], [33, 4.405405405],$

$[35, 4.435897436], [37, 4.463414634], [39, 4.488372093], [41, 4.511111111],$

$$[43, 4.531914894], \quad [45, 4.551020408], \quad [47, 4.568627451], \quad [49, 4.584905660],$$
$$[51, 4.600000000], \quad [53, 4.614035088], \quad [55, 4.627118644], \quad [57, 4.639344262],$$
$$[59, 4.650793651], \quad [61, 4.661538462], \quad [63, 4.671641791], \quad [65, 4.681159420],$$
$$[67, 4.690140845], \quad [69, 4.698630137], \quad [71, 4.706666667], \quad [73, 4.714285714],$$
$$[75, 4.721518987], \quad [77, 4.728395062], \quad [79, 4.734939759], \quad [81, 4.741176470],$$
$$[83, 4.747126437], \quad [85, 4.752808989], \quad [87, 4.758241758], \quad [89, 4.763440860],$$
$$[91, 4.768421052], \quad [93, 4.773195877], \quad [95, 4.777777778], \quad [97, 4.782178218],$$
$$[99, 4.786407767]$$

```
> C:=seq([n,c(n)],n=X);
```

$$C := [1, 1.600000000], \quad [3, 2.190476190], \quad [5, 2.755555556], \quad [7, 3.142857143],$$
$$[9, 3.418803419], \quad [11, 3.624242424], \quad [13, 3.782805430], \quad [15, 3.908771930],$$
$$[17, 4.011204481], \quad [19, 4.096109840], \quad [21, 4.167619048], \quad [23, 4.228663446],$$
$$[25, 4.281379310], \quad [27, 4.327359618], \quad [29, 4.367816092], \quad [31, 4.403686636],$$
$$[33, 4.435708435], \quad [35, 4.464468865], \quad [37, 4.490441661], \quad [39, 4.514013119],$$
$$[41, 4.535501355], \quad [43, 4.555170708], \quad [45, 4.573242630], \quad [47, 4.589904047],$$
$$[49, 4.605313823], \quad [51, 4.619607843], \quad [53, 4.632903013], \quad [55, 4.645300462],$$
$$[57, 4.656888122], \quad [59, 4.667742804], \quad [61, 4.677931905], \quad [63, 4.687514807],$$
$$[65, 4.696544035], \quad [67, 4.705066218], \quad [69, 4.713122891], \quad [71, 4.720751174],$$
$$[73, 4.727984344], \quad [75, 4.734852320], \quad [77, 4.741382075], \quad [79, 4.747597987],$$
$$[81, 4.753522149], \quad [83, 4.759174630], \quad [85, 4.764573695], \quad [87, 4.769736011],$$
$$[89, 4.774676815], \quad [91, 4.779410063], \quad [93, 4.783948565], \quad [95, 4.788304094],$$
$$[97, 4.792487496], \quad [99, 4.796508777]$$

Plot the Sequences

Now we plot the sequences for the indexing set we have defined. We set the n-range at 0 to nmax. From our numerical calculations above, we determine a common y-range for the sequences. Using the graphs, one can make a conjecture as to the limit of each sequence. We save each plot so that the 3 graphs can be displayed simultaneously at the end.

```
> seqrange:=-1..5;
```

$$seqrange := -1 \ldots 5$$

```
> plot([A],x=0..nmax,y=seqrange,style=point,symbol=DIAMOND);
```

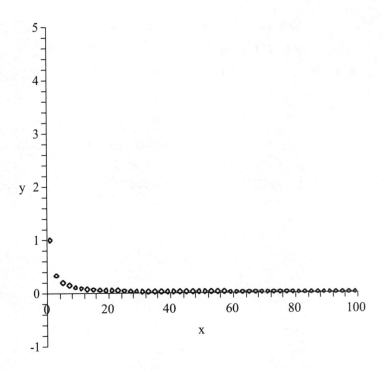

```
> gran:=plot([A],x=0..nmax,y=seqrange,style=point,symbol=DIAMOND):
> plot([B],n=0..nmax,y=seqrange,style=point,symbol=BOX,color=green);
```

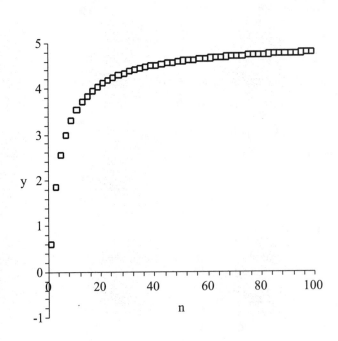

```
> grbn:=plot([B],n=0..nmax,y=seqrange,style=point,symbol=BOX,color=green):
> plot([C],n=0..nmax,y=seqrange,symbol=CROSS,style=point,color=NAVY);
```

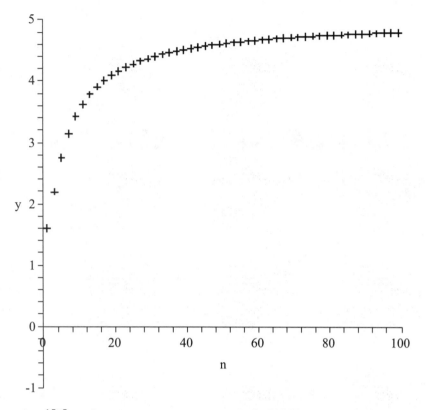

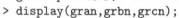

```
> grcn:=plot([C],n=0..nmax,y=seqrange,symbol=CROSS,style=point,color=NAVY):
> display(gran,grbn,grcn);
```

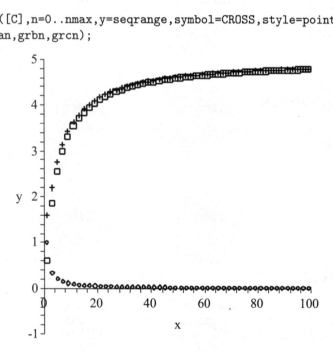

```
> restart;
```

Visual Guide for Lab 7

Our goal in this exploration is to use the power of graphics to understand properties of sequences. You should think of the code in this worksheet as a template. You should feel free to modify the code as each individual example suggests. For this lab, we want to be able to graph a sequence and compute the terms for large values of n. Examining the graph should help you answer the questions in the lab.

Load the Plotting Packages

```
> with(plots):
```

Define the Sequence

In the first step of this section of the code we define the sequence that we are investigating. The first sequence is $(a(n)) = (1/n)$.

```
> a:=n->1./n;
```

$$a := n \mapsto 1.0\,\frac{1}{n}$$

Enter the Indexing Set

Remember that the indexing set is just a subset of the domain of the sequence. Set the maximum value for the indexing set. We have initially set this max at 500. We count by fives and start the indexing set at 1. These numbers may be changed, if you find the need or have the desire.

```
> nmax:=500:
> X:=1:
> for i from 6 to nmax by 5 do
> X:=X,i;
> od:
```

Computing Some Terms of the Sequence

```
> A:=seq([n,a(n)],n=X);
```

$A := [1, 1.0], [6, 0.1666666667], [11, 0.09090909091], [16, 0.06250000000], [21, 0.04761904762],$

$[26, 0.03846153846], [31, 0.03225806452], [36, 0.02777777778], [41, 0.02439024390],$

$[46, 0.02173913044], [51, 0.01960784314], [56, 0.01785714286], [61, 0.01639344262],$

$[66, 0.01515151515], [71, 0.01408450704], [76, 0.01315789474], [81, 0.01234567901],$

$[86, 0.01162790698], [91, 0.01098901099], [96, 0.01041666667], [101, 0.009900990099],$

$[106, 0.009433962264], [111, 0.009009009009], [116, 0.008620689655], [121, 0.008264462810],$

$[126, 0.007936507936], [131, 0.007633587786], [136, 0.007352941176], [141, 0.007092198582],$

$[146, 0.006849315068], [151, 0.006622516556], [156, 0.006410256410], [161, 0.006211180124],$

$[166, 0.006024096386], [171, 0.005847953216], [176, 0.005681818182], [181, 0.005524861878],$

$[186, 0.005376344086], [191, 0.005235602094], [196, 0.005102040816], [201, 0.004975124378],$

$[206, 0.004854368932], [211, 0.004739336493], [216, 0.004629629630], [221, 0.004524886878],$

$[226, 0.004424778761], [231, 0.004329004329], [236, 0.004237288136], [241, 0.004149377593],$

$[246, 0.004065040650], [251, 0.003984063745], [256, 0.003906250000], [261, 0.003831417624],$

$[266, 0.003759398496], [271, 0.003690036900], [276, 0.003623188406], [281, 0.003558718861],$

$[286, 0.003496503496], [291, 0.003436426117], [296, 0.003378378378], [301, 0.003322259136],$

$[306, 0.003267973856], [311, 0.003215434084], [316, 0.003164556962], [321, 0.003115264798],$

$[326, 0.003067484663], [331, 0.003021148036], [336, 0.002976190476], [341, 0.002932551320],$

$[346, 0.002890173410], [351, 0.002849002849], [356, 0.002808988764], [361, 0.002770083102],$

$[366, 0.002732240437], [371, 0.002695417790], [376, 0.002659574468], [381, 0.002624671916],$

$[386, 0.002590673575], [391, 0.002557544757], [396, 0.002525252525], [401, 0.002493765586],$

$[406, 0.002463054187], [411, 0.002433090024], [416, 0.002403846154], [421, 0.002375296912],$

$[426, 0.002347417840], [431, 0.002320185615], [436, 0.002293577982], [441, 0.002267573696],$

$[446, 0.002242152466], [451, 0.002217294900], [456, 0.002192982456], [461, 0.002169197397],$

$[466, 0.002145922747], [471, 0.002123142250], [476, 0.002100840336], [481, 0.002079002079],$

$[486, 0.002057613169], [491, 0.002036659878], [496, 0.002016129032]$

Plot the Sequence

Now we plot the sequence for the indexing set we have defined. We set the x-range at 0 to xmax. From our numerical calculations above, we determine the y-range. Using the graph and numerical calculations, answer the questions in the lab.

```
> plot([A],n=0..nmax,y=-.2..0.2,style=point,symbol=DIAMOND);
```

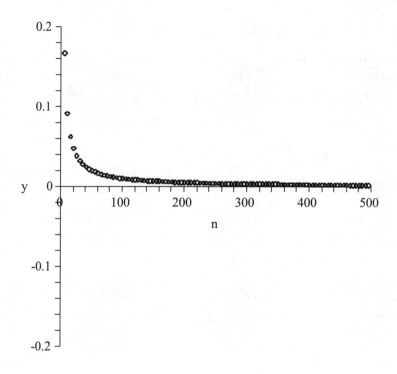

Computing Differences in Terms

If you are feeling lazy, you can have Maple compute the absolute value of the differences of some terms. Adjust the n values as you please.

```
> abs(a(50)-a(25));
```

$$0.02000000000$$

```
> abs(a(100)-a(80));
```

$$0.00250000000$$

```
> abs(a(200)-a(150));
```

$$0.001666666667$$

```
> abs(a(500)-a(400));
```

$$0.000500000000$$

```
> restart;
```

Visual Guide for Lab 9

Our goal is to determine a definition of the continuity of a function $f(x)$ at a domain point x_0 based upon sequences. Let $(x(n))$ represent a sequence which converges to x_0. The goal is to determine how the sequence $(f(x(n)))$ behaves when f is continuous at x_0.

Load the Plotting Package

```
> with(plots):
```

Defining the Function

First we define the function.

```
> f:=x->x^2-1.;
```

$$f := x \mapsto x^2 - 1.0$$

We include an example of a piecewise-defined function.

```
> g:=x->piecewise(x<4,x-2,x=4,12,x>4,6-2*x);
```

$$g := x \mapsto \begin{cases} x - 2 & x < 4 \\ 12 & x = 4 \\ 6 - 2x & 4 < x \end{cases}$$

Now we name the point at which we wish to investigate continuity.

```
> x0:=0;
```

$$x0 := 0$$

We compute the function value at this point.

```
> f(x0);
```

$$-1.0$$

Using Your Previous Knowledge To Decide If the Function Is Continuous at the Desired Point and If the Limit of $f(x)$ as x approaches x_0 exists

First we graph the function near x_0. (We include an example to show you how to deal with piecewise-defined functions.) Do you think f is continuous at x_0? Do you think $\lim_{x \to x_0} f(x)$ exists?

```
> plot(f(x),x=-2..2);
> fplot:=plot(f(x),x=-2..2):
```

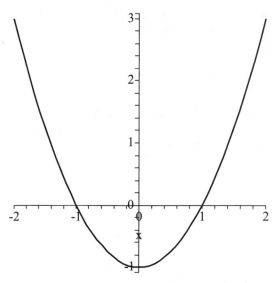

What follows are the directions for plotting a piecewise-defined function. There are two parts. The first part is the graph of the function up to the point where the definition splits and the second part is the graph after the point where the definition splits.

```
> pt1:=plot(g,0..4,style=point):
  pt2:=plot(g,4..6,style=point):
  display([pt1,pt2]);
```

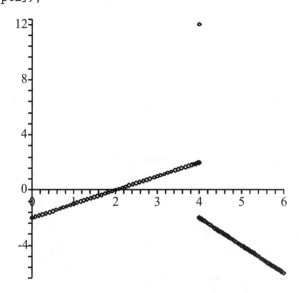

As an alternate approach we calculate a table of values. Do you think f is continuous at x_0? Do you think $\lim_{x \to x_0} f(x)$ exists?

```
> print("x","f(x)");
> for i from 1 to 5 do
> print(x0-(10.^(-i)),f(x0-(10.^(-i))) );
> od;
> print(x0,f(x0));
> for i from 1 to 5 do
> print(x0+(10.^(i-6)),f(x0+(10.^(i-6))) );
> od;
```

$$\text{``}x\text{''},\ \text{``}f(x)\text{''}$$
$$-0.1000000000,\ -0.9900000000$$
$$-0.01000000000,\ -0.9999000000$$
$$-0.001000000000,\ -0.9999990000$$
$$-0.0001000000000,\ -0.9999999900$$
$$-0.00001000000000,\ -0.9999999999$$
$$0,\ -1.0$$
$$0.00001000000000,\ -0.9999999999$$
$$0.0001000000000,\ -0.9999999900$$
$$0.001000000000,\ -0.9999990000$$
$$0.01000000000,\ -0.9999000000$$
$$0.1000000000,\ -0.9900000000$$

Defining the sequences that converge to x_0

```
> a:=x->1./x;
```

$$a := x \mapsto 1.0\,x^{-1}$$

```
> b:=x->x/(x^2+1.);
```

$$b := x \mapsto \frac{x}{x^2 + 1.0}$$

Computing Terms of Sequences and Finding the Limits of Sequences of Outputs $(f(a(n))$ and $(f(b(n))$

We define the indexing set for the sequences $(a(n))$, $(b(n))$, $(f(a(n)))$, and $(f(b(n)))$. Set the maximum value for the indexing set. We have initially set this max at 200. We count by fives and start the indexing set at 1. These numbers may be changed, if you find the need or have the desire.

```
> xmax:=200:
> X:=1:
> for i from 6 to xmax by 5 do
> X:=X,i;
> od:
```

First a numerical approach. We compute some terms of the sequences and look for patterns. Does $(a(n))$ converge to x_0? Does $(b(n))$ converge to x_0? Does $(f(a(n)))$ converge? Does $(f(b(n)))$ converge?

```
> A:=seq([x,a(x)],x=X);
```

$A := [1, 1.0], [6, 0.1666666667], [11, 0.09090909091], [16, 0.06250000000], [21, 0.04761904762],$
$\qquad [26, 0.03846153846], [31, 0.03225806452], [36, 0.02777777778], [41, 0.02439024390],$
$\qquad [46, 0.02173913044], [51, 0.01960784314], [56, 0.01785714286], [61, 0.01639344262],$
$\qquad [66, 0.01515151515], [71, 0.01408450704], [76, 0.01315789474], [81, 0.01234567901],$
$\qquad [86, 0.01162790698], [91, 0.01098901099], [96, 0.01041666667], [101, 0.009900990099],$
$\qquad [106, 0.009433962264], [111, 0.009009009009], [116, 0.008620689655], [121, 0.008264462810],$
$\qquad [126, 0.007936507936], [131, 0.007633587786], [136, 0.007352941176], [141, 0.007092198582],$
$\qquad [146, 0.006849315068], [151, 0.006622516556], [156, 0.006410256410], [161, 0.006211180124],$
$\qquad [166, 0.006024096386], [171, 0.005847953216], [176, 0.005681818182], [181, 0.005524861878],$
$\qquad [186, 0.005376344086], [191, 0.005235602094], [196, 0.005102040816]$

```
> B:=seq([x,b(x)],x=X);
```

$B := [1, 0.5000000000], [6, 0.1621621622], [11, 0.09016393443], [16, 0.06225680934],$
$\qquad [21, 0.04751131222], [26, 0.03840472674], [31, 0.03222453222], [36, 0.02775636083],$
$\qquad [41, 0.02437574316], [46, 0.02172886160], [51, 0.01960030746], [56, 0.01785145043],$
$\qquad [61, 0.01638903815], [66, 0.01514803764], [71, 0.01408171361], [76, 0.01315561710],$
$\qquad [81, 0.01234379762], [86, 0.01162633500], [91, 0.01098768413], [96, 0.01041553651],$
$\qquad [101, 0.009900019604], [106, 0.009433122720], [111, 0.009008277877], [116, 0.008620049045],$
$\qquad [121, 0.008263898375], [126, 0.007936008062], [131, 0.007633142990], [136, 0.007352543656],$
$\qquad [141, 0.007091841867], [146, 0.006848993761], [151, 0.006622226121], [156, 0.006409993015],$
$\qquad [161, 0.006210940514], [166, 0.006023877781], [171, 0.005847753232], [176, 0.005681634761],$
$\qquad [181, 0.005524693242], [186, 0.005376188687], [191, 0.005235458582], [196, 0.005101908009]$

```
> FA:=seq([x,f(a(x))],x=X);
```

$FA := [1, 0.0], [6, -0.9722222222], [11, -0.9917355372], [16, -0.9960937500], [21, -0.9977324263],$
$\qquad [26, -0.9985207101], [31, -0.9989594173], [36, -0.9992283951], [41, -0.9994051160],$
$\qquad [46, -0.9995274102], [51, -0.9996155325], [56, -0.9996811224], [61, -0.9997312550],$
$\qquad [66, -0.9997704316], [71, -0.9998016267], [76, -0.9998268698], [81, -0.9998475842],$
$\qquad [86, -0.9998647918], [91, -0.9998792416], [96, -0.9998914931], [101, -0.9999019704],$
$\qquad [106, -0.9999110004], [111, -0.9999188378], [116, -0.9999256837], [121, -0.9999316987],$
$\qquad [126, -0.9999370118], [131, -0.9999417283], [136, -0.9999459343], [141, -0.9999497007],$
$\qquad [146, -0.9999530869], [151, -0.9999561423], [156, -0.9999589086], [161, -0.9999614212],$
$\qquad [166, -0.9999637103], [171, -0.9999658014], [176, -0.9999677169], [181, -0.9999694759],$
$\qquad [186, -0.9999710949], [191, -0.9999725885], [196, -0.9999739692]$

```
> FB:=seq([x,f(b(x))],x=X);
```

$FB := [1, -0.7500000000], [6, -0.9737034332], [11, -0.9918704649], [16, -0.9961240897],$

$\qquad [21, -0.9977426752], [26, -0.9985250770], [31, -0.9989615795], [36, -0.9992295844],$

$\qquad [41, -0.9994058231], [46, -0.9995278566], [51, -0.9996158279], [56, -0.9996813257],$

$\qquad [61, -0.9997313994], [66, -0.9997705370], [71, -0.9998017053], [76, -0.9998269297],$

$\qquad [81, -0.9998476307], [86, -0.9998648283], [91, -0.9998792708], [96, -0.9998915166],$

$\qquad [101, -0.9999019896], [106, -0.9999110162], [111, -0.9999188509], [116, -0.9999256948],$

$\qquad [121, -0.9999317080], [126, -0.9999370198], [131, -0.9999417351], [136, -0.9999459401],$

$\qquad [141, -0.9999497058], [146, -0.9999530913], [151, -0.9999561461], [156, -0.9999589120],$

$\qquad [161, -0.9999614242], [166, -0.9999637129], [171, -0.9999658038], [176, -0.9999677190],$

$\qquad [181, -0.9999694778], [186, -0.9999710966], [191, -0.9999725900], [196, -0.9999739705]$

Hopefully, you see a pattern. To reinforce your feeling, we will also look at the graphs of the sequences $(f(a(n)))$ and $(f(b(n)))$. Be careful—what follows is not a graph of $f(x)$, but of the sequences $(f(a(n)))$ and $(f(b(n)))$.

```
> plot([FA],x=0..xmax, y=-1.2..-.8, style =point, color=blue);
```

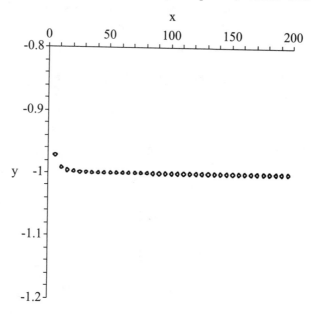

```
> plot([FB],x=0..xmax, y=-1.2..-.8, style =point, color=black);
```

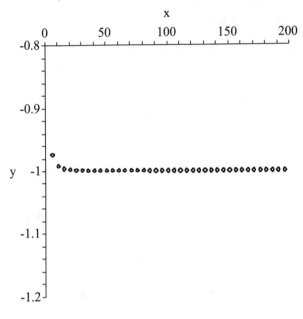

Now we also look at the graphs of the sets of points
$$\{(a(n), f(a(n))) : n \text{ in } X\} \quad \text{and} \quad \{(b(n), f(b(n))) : n \text{ in } X\}.$$
These graphs are "subgraphs" of the graph of $f(x)$. First we will show the plot of the sequence and then the plot of the sequence superimposed on the plot of the function.

```
> plot({seq([a(x),f(a(x))],x=X)},x=-1..1, y=-1.2..-.8, style =point,
  color=blue);
> seqaplot:=plot({seq([a(x),f(a(x))],x=X)},x=-1..1, y=-1.2..-.8,
  style =point, color=blue):
```

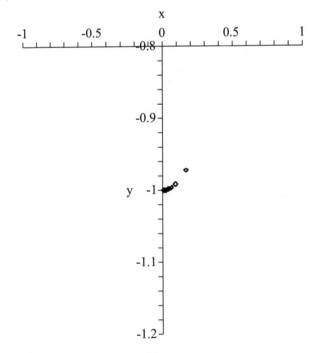

> display({fplot,seqaplot});

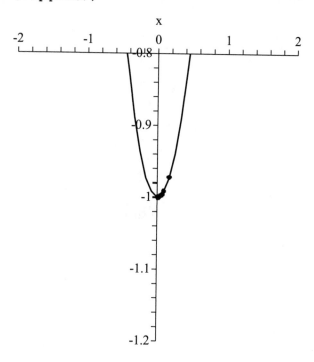

> plot({seq([b(x),f(b(x))],x=X)},x=-1..1, y=-1.2..-.7, style =point,
  color=black);
> seqbplot:=plot({seq([b(x),f(b(x))],x=X)},x=-1..1, y=-1.2..-.7,
  style =point, color=black):

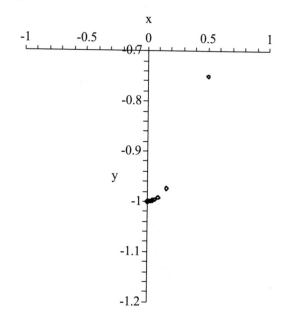

> display({seqbplot,fplot});

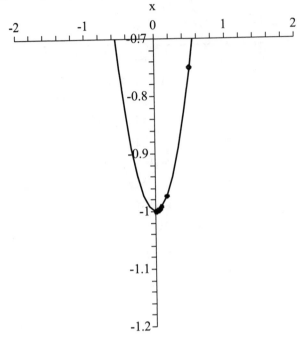

> restart;

Visual Guide for Lab 10

Our goal is to determine a delta-epsilon definition of continuity at a point x_0. Loosely speaking, we will do this by looking at an arbitrary horizontal band centered about $f(x_0)$. The goal is to determine if there is a corresponding vertical band centered about x_0 such that domain values in the vertical band map to function values in the horizontal band.

Load the Plotting Package

```
> with(plots):
```

Defining the Function

First we define the function.

```
> f:=x->abs(x);
```

$$f := \text{abs}$$

We include an example of a piecewise-defined function.

```
> g:=x->piecewise(x<4,(x^2-16)/(x-4),x=4,7,x>4,(x^2-16)/(x-4));
```

$$g := x \mapsto \begin{cases} \frac{x^2-16}{-4+x} & x < 4 \\ 7 & x = 4 \\ \frac{x^2-16}{-4+x} & 4 < x \end{cases}$$

Now we name the point at which we wish to investigate continuity.

```
> x0:=0;
```

$$x0 := 0$$

We compute the function value at this point.

```
> f(x0);
```

$$0$$

Graphing the Function

Use the graph below or some other method to decide if the function is continuous at x_0. We include an example of plotting a piecewise-defined function. You can change the range on the domain values if you want or need to.

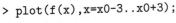

```
> plot(f(x),x=x0-3..x0+3);
```

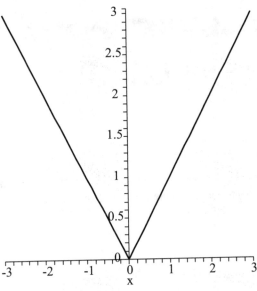

What follows are the directions for plotting a piecewise-defined function. There are two parts. The first part is the graph of the function up to the point where the definition splits and the second part is the graph after the point where the definition splits.

```
> pt1:=plot(g,0..4,style=point):
  pt2:=plot(g,4..6,style=point):
  display([pt1,pt2]);
```

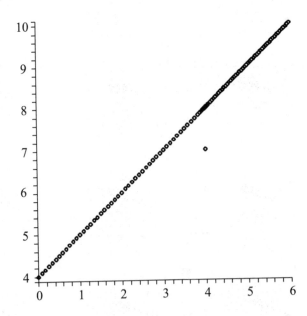

Setting up the Horizontal epsilon-band

We set the width of the horizontal band.

> `epsilon:=1;`

$$\epsilon := 1$$

Plotting the Horizontal Band

We define an interval centered at x_0 as the domain of the graph. For the radius of the interval, the value 2 was chosen arbitrarily. You may need to adjust it.

> `xrange:=x0-2..x0+2;`

$$xrange := -2\ldots 2$$

Now we will plot the function and the horizontal band.

> `plot({f(x),f(x0)-epsilon,f(x0)+epsilon},x=xrange,color=green);`

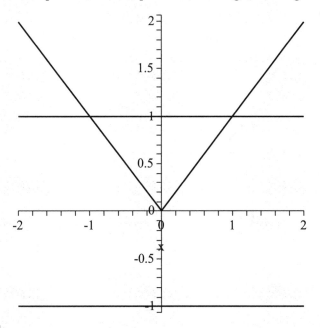

Save the above graph.

> `gra:=plot({f(x),f(x0)-epsilon,f(x0)+epsilon},x=xrange,color=green):`

Determining the Vertical Band

Now we will show you how to experiment with different vertical lines. Suppose our guess for the width of the band centered about x_0 was 1.5. Then we would graph two vertical lines as shown below. We need to set some y limits. You can determine reasonable values for these from your graph above.

> `implicitplot({x=x0-1.5,x=x0+1.5},x=xrange,y=-2..2);`

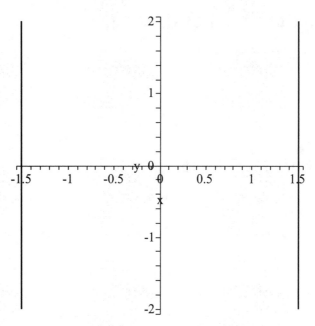

```
> guess:=implicitplot({x=x0-1.5,x=x0+1.5},x=xrange,y=-2..2):
```

Now we add the vertical lines to the graph of the function and the horizontal band.

```
> display({guess,gra});
```

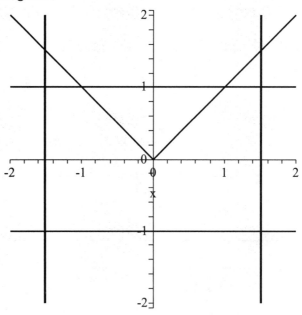

This was a bad guess. Did you expect us to give you a good one?!

```
> restart;
```

Visual Guide for Lab 12

In this exercise, we are examining two ideas. First, we want to understand the limit of a sequence of functions. Then we will examine the two kinds of convergence that are possible when talking about sequences of functions.

Load the Plotting Package

Load the plotting package.

```
> with(plots):
```

12.1 Example 1

In this example, we consider the functions (x^n) on the common domain $[0, 1]$. We want to understand graphically the pointwise limit of this sequence. We will graph some of the functions in this sequence and then focus on the value of the limit function at a particular point in the domain.

We begin by defining the sequence of functions. We enter the formula for the first 300 functions in the sequence. The number 300 was chosen arbitrarily. We let $f[n](x) = x^n$.

```
> P:=x:
> for k from 2 to 300 do
> P:= P,x^k:
> od:
> for n from 1 to 300 do f[n]:=unapply(P[n],x):
> od:
```

Now we plot the functions as requested in question 1a. Here we graph some of the functions in the sequence (x^n).

```
> plot({f[1](x),f[3](x),f[5](x),f[10](x),f[15](x)},x=0..1);
```

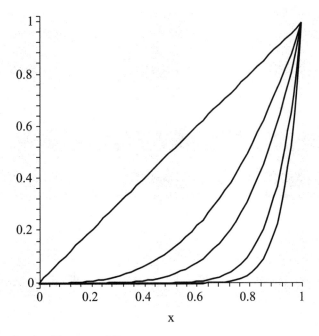

We save the above graph, should we need it.

> sampf:=%:

In question 1b, we want to investigate the behavior of the sequence of functions when they are evaluated at $x_0 = .5$. We specify that point in the next line.

> x0:=.5;

$$x0 := 0.5$$

Now we will evaluate the first six functions in the sequence at the point $x = x_0$.

> for n from 1 to 6 do
> f[n](x0);
> od;

$$0.5$$
$$0.25$$
$$0.125$$
$$0.0625$$
$$0.03125$$
$$0.015625$$

Below we plot the points $(x_0, f[n](x_0))$ for $n = 1$ to 6. Notice they all lie on the vertical line $x = x_0$, as expected.

> pointplot({seq([x0,f[n](x0)],n=1..6)});

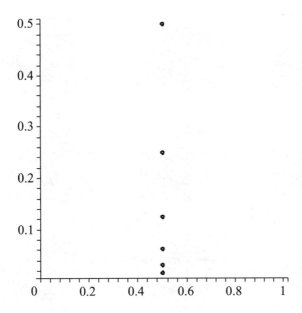

We save the above graph.

> b:=%:

We would like to look at those points as part of the graphs of the functions $f[n](x)$. We first graph the first 6 functions in the sequence.

> plot({seq(f[n](x),n=1..6)},x=0..1);

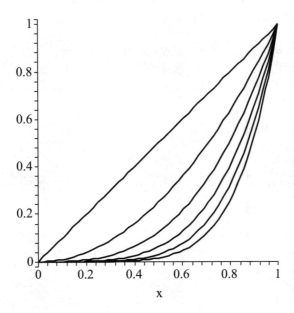

> c:=%:

We will display the first six points in the sequence $(f[n](x_0))$ as points on the corresponding functions.

> display(b,c);

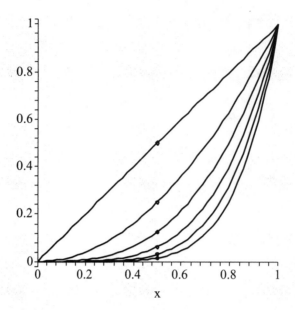

What does the limit of those function values at $x = x_0$ appear to be? Repeat this process for other values of x in the domain $[0, 1]$. Adjust the above code as needed. We can define the pointwise limit of this sequence of functions. We do that below and then plot the pointwise limit. Notice that style = point was chosen so that you would get an accurate graph.

```
> flim := x -> piecewise(x=0,0,x>0 and x<1,0,x=1,1);
```

$$flim := x \mapsto \begin{cases} 0 & x = 0 \\ 0 & -x < 0 \text{ and } x < 1 \\ 1 & x = 1 \end{cases}$$

```
> plot(flim(x),x=0..1,discont=true,style = point, color = khaki);
```

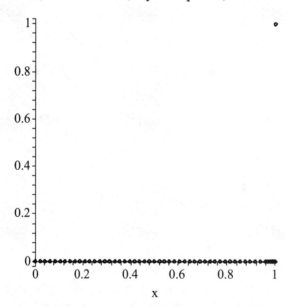

```
> grflim:=%:
```

Example 2

We repeat the process we performed above using the sequence of functions $((x^n)/(1+x^n))$ with common domain $[0, 2]$. We will graph some of the functions in this sequence and then focus on the value of the limit function at a particular point in the domain.

We begin by defining the sequence of functions. We enter the formula for the first 300 functions in the sequence. The number 300 was chosen arbitrarily. We let $g[n](x) = (x^n)/(1+x^n)$.

```
> P:=x/(1+x):
> for k from 2 to 300 do
> P:= P,(x^k)/(1+x^k):
> od:
> for n from 1 to 300 do g[n]:=unapply(P[n],x):
> od:
```

Now we plot the functions as requested in question 2a.

```
> plot({g[1](x),g[10](x),g[50](x),g[100](x),g[300](x)},x=0..2);
```

Here we graph some of the functions in the sequence $((x^n)/(1+x^n))$.

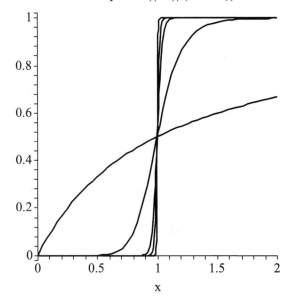

```
> sampg:=%:
```

In question 2b, we want to investigate the behavior of the sequence of functions when they are evaluated at $x_0 = .5$. We specify that point in the next line.

```
> x0:=.5;
```

$$x0 := 0.5$$

Now we will evaluate the first six functions in the sequence at the point $x = x_0$.

```
> for n from 1 to 6 do
> g[n](x0);
> od;
```

$$0.3333333333$$
$$0.2000000000$$
$$0.1111111111$$

$$0.05882352941$$
$$0.03030303030$$
$$0.01538461538$$

Below we plot the points $(x_0, g[n](x_0))$ for $n = 1$ to 6. Notice they all lie on the vertical line $x = x_0$, as expected.

```
> pointplot({seq([x0,g[n](x0)],n=1..6)});
```

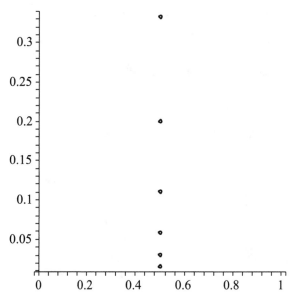

We save the above graph.

```
> b:=%:
```

We would like to look at those points as part of the graphs of the functions $g[n](x)$. We first graph the first 6 functions in the sequence.

```
> plot({seq(g[n](x),n=1..6)},x=0..2);
```

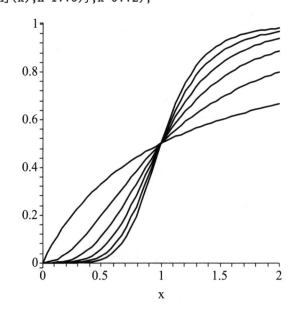

```
> c:=%:
```

We will display the first six points in the sequence $(g[n](x_0))$ as points on the corresponding functions.

```
> display(c,b);
```

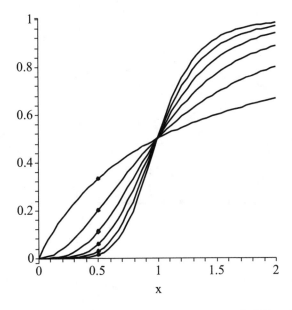

What does the limit of those function values at $x = x_0$ appear to be? Repeat this process for other values of x in the domain $[0, 2]$. Adjust the above code as needed. We can define the pointwise limit of this sequence of functions. We do that below and then plot the pointwise limit.

```
> glim:=x->piecewise(x=0,0, x<1,0,x=1,.5,1<x and x<2,1);
```

As before, after additional evaluations, define the pointwise limit of the sequence of functions.

$$glim := x \mapsto \begin{cases} 0 & x = 0 \\ 0 & x < 1 \\ 0.5 & x = 1 \\ 1 & -x < -1 \text{ and } x < 2 \end{cases}$$

Evaluate the limit function at the interesting point $x = 1$.

```
> glim(1);
```

$$0.5$$

Now we plot the limit function.

```
> pt1:=plot(glim(x),x=0..1,style=point, color = khaki):
> pt2:=plot(glim(x),x=1..2,style=point, color=khaki):
> display(pt1,pt2);
```

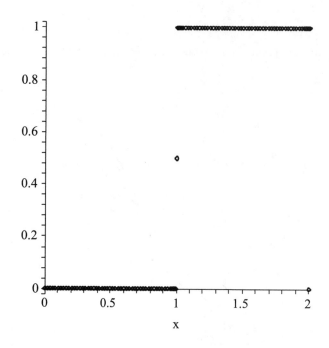

> grglim:=%:

Example 3

We repeat the process we performed above using the sequence of functions $((x)/(1 + nx^2))$ with common domain $[0, 1]$. We will graph some of the functions in this sequence and then focus on the value of the limit function at a particular point in the domain.

We begin by defining the sequence of functions. We enter the formula for the first 300 functions in the sequence. The number 300 was chosen arbitrarily. We let $h[h](x) = (x)/(1 + nx^2)$.

```
> P:=x/(1+x):
> for k from 2 to 300 do
> P:= P,(x)/(1+k*x^2):
> od:
> for n from 1 to 300 do h[n]:=unapply(P[n],x):
> od:
```

Now we plot the functions as requested in question 3a.

```
> plot({h[1](x),h[10](x),h[50](x),h[100](x),h[300](x)},x=0..1);
```

Here we graph some of the functions in the sequence $((x^n)/(1 + x^n))$.

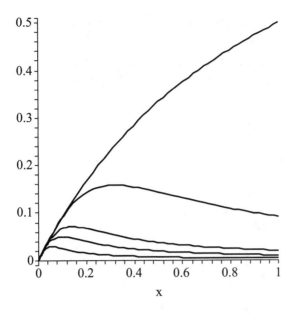

> samph:=%:

In question 3b, we want to investigate the behavior of the sequence of functions when they are evaluated at $x_0 = .5$. We specify that point in the next line.

> x0:=.5;

$$x0 := 0.5$$

Now we will evaluate the first twenty functions in the sequence at the point $x = x_0$.

> for n from 1 to 20 do
> h[n](x0);
> od;

0.3333333333
0.3333333333
0.2857142857
0.2500000000
0.2222222222
0.2000000000
0.1818181818
0.1666666667
0.1538461538
0.1428571429
0.1333333333
0.1250000000
0.1176470588
0.1111111111
0.1052631579
0.1000000000
0.09523809524
0.09090909091
0.08695652174
0.08333333333

Below we plot the points $(x_0, h[n](x_0))$ for $n = 1$ to 20. Notice they all lie on the vertical line $x = x_0$, as expected.

```
> pointplot({seq([x0,h[n](x0)],n=1..20)});
```

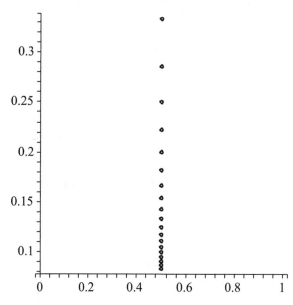

We save the above graph.

```
> b:=%:
```

We would like to look at those points as part of the graphs of the functions $h[n](x)$. We first graph the first 20 functions in the sequence.

```
> plot({seq(h[n](x),n=1..20)},x=0..1);
```

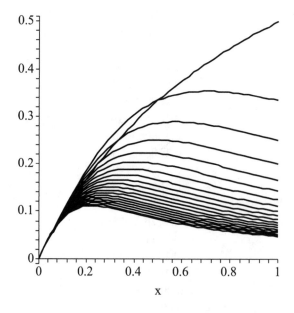

```
> c:=%:
```

We will display the first 20 points in the sequence $(h[n](x_0))$ as points on the corresponding functions.

```
> display(c,b);
```

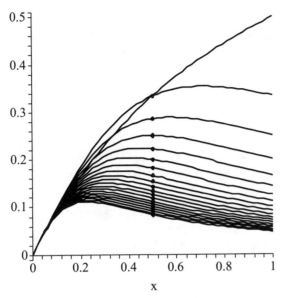

What does the limit of those function values at $x = x_0$ appear to be? Repeat this process for other values of x in the domain $[0, 1]$. Adjust the above code as needed. We can define the pointwise limit of this sequence of functions. We do that below and then plot the pointwise limit.

```
> hlim:=x->0;
```

As before, after additional evaluations, define the pointwise limit of the sequence of functions.

$$hlim := 0$$

Now we plot the limit function.

```
> plot(hlim(x),x=0..1,style = point, color=khaki);
```

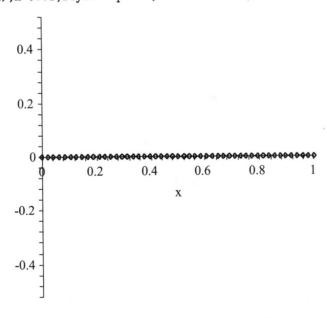

```
> grhlim:=%:
```

12.2 Example 1 revisited

Now that we can compute the pointwise limit of a sequence of functions, we want to examine the convergence from a more global view. Do the sequences of function values converge rather uniformly, regardless of the domain values, or does the rate of convergence seem to depend upon the domain value? We will form an epsilon band about the limit function. We let "ep" represent epsilon. In the first case, epsilon is .5.

```
> ep:=.5;
```

$$ep := 0.5$$

Since we have already saved the graph from question 1a of Section 2 of the lab as sampf and the graph of the limit function as "grflim," we need only graph the functions $f+$ epsilon and $f-$ epsilon. Then we will display all the graphs.

```
> plot({flim(x)+ep,flim(x)-ep},x=0..1,style=point, color= brown);
```

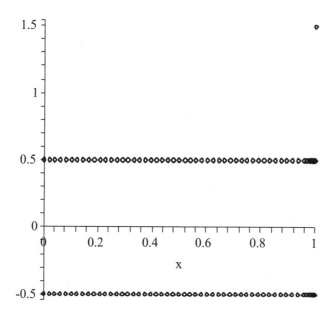

```
> fband:=%:
> display(sampf, grflim, fband);
```

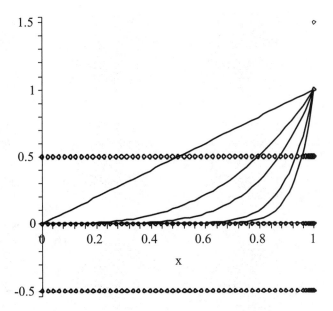

Look at how the terms of the first sequence of functions approach the limit function. Remember the limit function sits on the x-axis except at the point $x = 1$, where the value is 1. Is there a point in the indexing set for the sequence of functions, after which the sequence functions fall entirely within the epsilon band about the limit function?

To experiment with graphing different functions $f[n](x)$ in the epsilon band of the limit function, use the code below.

```
> plot(f[40](x),x=0..1);
> d:=%:
> display(d,grflim, fband);
```

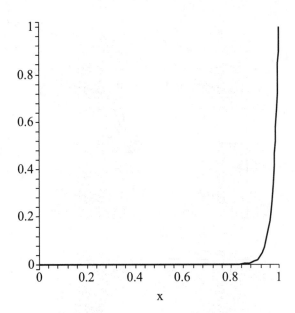

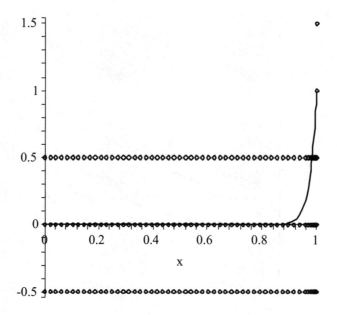

Example 2 revisited

We consider the same questions for the second sequence of functions. We will form an epsilon band about the limit function. We let "ep" represent epsilon. In the first case, epsilon is .3.

```
> ep:=.3;
```

$$ep := 0.3$$

Since we want to adjust the domain interval we will graph again the functions from question 2 in Section 2 of the lab as as well as the pointwise limit. Then we will display all the graphs.

```
> plot({g[1](x),g[10](x),g[50](x),g[100](x),g[300](x)},x=.95..1.05);
```

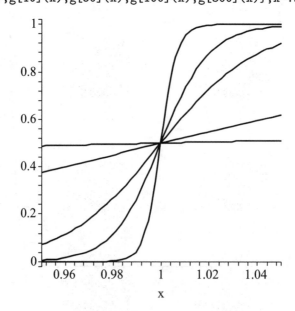

```
> sampg2:=%:
> part1:=plot(glim(x),x=.95..1,style=point, color = khaki):
> part2:=plot(glim(x),x=1..1.05,style=point, color = khaki):
> display(part1,part2):
> glim2:=%:
> plot({glim(x)+ep,glim(x)-ep},x=.95..1.05,style=point, color= brown):
> gband:=%:
> display(sampg2, gband,glim2);
```

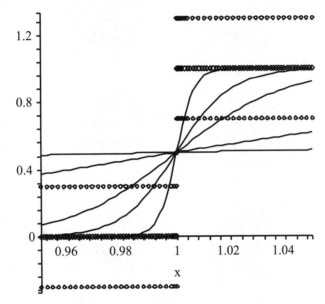

To experiment with graphing different functions $g[n](x)$ in the epsilon band of the limit function, use the code below.

```
> plot(g[40](x),x=.95..1.05);
> d:=%:
> display(d, gband,glim2);
```

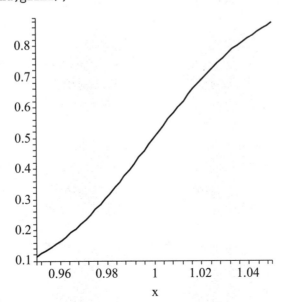

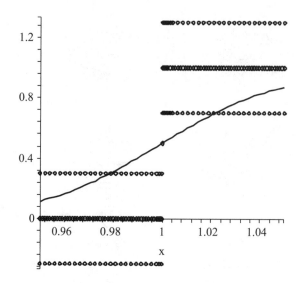

12.3   Example 3 revisited

We consider the same questions for the third sequence of functions. We will form an epsilon band about the limit function. We let "ep" represent epsilon. In the first case, epsilon is .3.

```
> ep:=.3;
```

$$ep := 0.3$$

Since we have already saved the graph from question 3a of Section 2 of the lab as samph and the graph of the limit function as grhlim, we need only graph the functions $h+$ epsilon and $h-$ epsilon. Then we will display all the graphs.

```
> plot({hlim(x)+ep,hlim(x)-ep},x=0..1,style=point, color= brown);
```

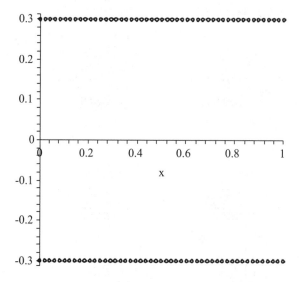

```
> hband:=%:
> display(samph, grhlim, hband);
```

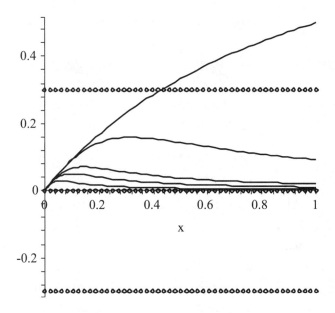

Look at how the terms of the first sequence of functions approach the limit function. Remember the limit function sits on the x-axis except at the point $x = 1$, where the value is 1. Is there a point in the indexing set for the sequence of functions, after which the sequence functions fall entirely within the epsilon band about the limit function?

To experiment with graphing different functions $h[n](x)$ in the epsilon band of the limit function, use the code below.

```
> plot(h[40](x),x=0..1);
> d:=%:
> display(d,grhlim, hband);
```

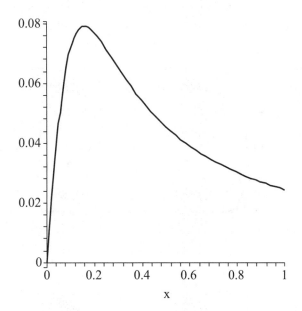

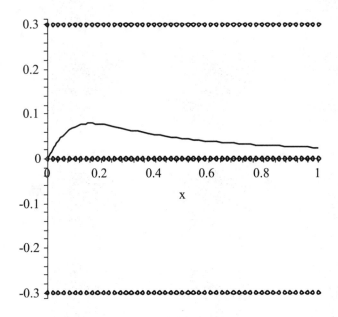

12.4 Example 4

We consider one more sequence of functions, which exhibits the same nice behavior you observed for Example 3. The sequence of functions is $(n * x * (exp((-n^2) * x)))$ on the domain $[0, 1]$.

```
> plot({x*exp(-x),2*x*exp(-4*x),3*x*exp(-9*x),4*x*exp(-16*x),5*x*exp(-25*x),
  6*x*exp(-36*x),10*x*exp(-100*x),11*x*exp(-121*x),12*x*exp(-144*x)},x=0..1);
```

Plot some terms in the sequence and try to determine the limit function.

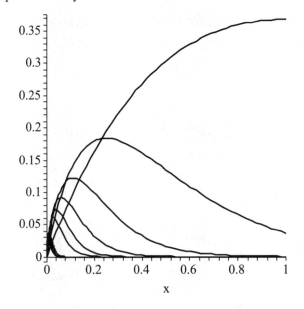

```
> sampk:=%:
> klim:=x->0;
```

Define the limit function.

$$klim := 0$$

```
> plot({klim(x),klim(x)+.1,klim(x)-.1},x=0..1);
> e:=%:
> display(e,sampk);
```

Graph some of the terms of the sequence of functions, the limit function, and the epsilon band.

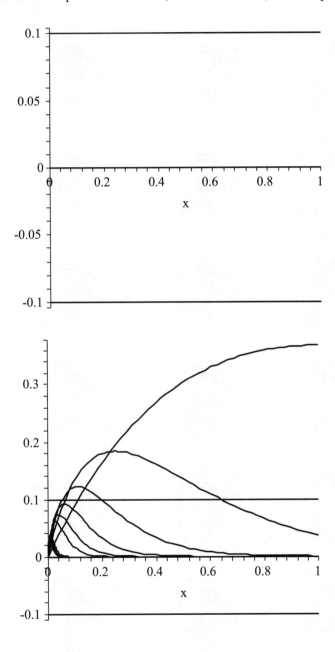

About the Authors

Joanne Snow is currently Professor of Mathematics at Saint Mary's College, Notre Dame, Indiana. She has been on the faculty there since 1983. In 1993, Dr. Snow was given the Spes Unica Award, an award for outstanding teaching. She has long had an interest in the use of writing to learn mathematics. Dr. Snow has written and lectured on the topic. Because of this interest in writing, she has served on the college's Steering Committee for the Writing Proficiency Program and as a Co-Director of the program. Dr. Snow received her BA degree in mathematics from Loyola College, Baltimore, Maryland and her MS and PhD degrees in mathematics from the University of Notre Dame. Dr. Snow's scholarly interests now include the history of mathematics and pedagogy. Her interests in the history of mathematics include the development of the real number system and the life and work of Marston Morse. She is a member of MAA, CSHPM, and AWM. She has an article in each of the MAA publications: *Using Writing to Teach Mathematics* and *Environmental Mathematics in the Classroom.*

Kirk Weller is currently Associate Professor of Mathematics Education at the University of North Texas. Before joining the UNT mathematics faculty in 2002, Dr. Weller was a member of the faculty at Bethel College (IN), where he was named professor of the year in 1998. Dr. Weller received his BA degree in mathematics from Hope College and his MS and PhD degrees in mathematics from the University of Notre Dame. Dr. Weller is co-author of *Learning Linear Algebra with ISETL,* which is available online. Current research interests include the study of students' conceptions of mathematical infinity. Dr. Weller is a member of RUMEC (Research in Undergraduate Mathematics Education Community) and SIGMAA on RUME (Special Interest Group of the MAA on Research in Undergraduate Mathematics Education).